SHARK 4
The Encyclopedic

Summer 2002
Edited by Lytle Shaw and Emilie Clark

SHARK is a journal of poetics
and art writing.

The editors wish to thank Nina Katchadourian, Andrew Clark, Catharine Clark, Paul Chan, Jimbo Blachly, Matt Mullican, Allyson Strafella, and Bernadette Mayer. We would also like to thank the galleries, Derek Eller, Pierogi, D'Amelio Terras, Debs & Co, Murray Guy and Postmasters, whose artists contributed work to this issue.

Part of this issue was made possible by a generous contribution from the Fund for Poetry. Many thanks.

This issue is in memory of Maxine Hoffer, 1916-2002.

Shark was printed by Thomson-Shore. Design and book artwork by Emilie Clark. Cover artwork from Diderot's *L'Encyclopédie*.

Shark is published annually. Individual issues are $10. Subscriptions (2 issues) are $18 ($23 institutions). North American subscriptions outside of the US, add $3 for postage; other foreign subscriptions, add $5. Please make checks payable (in US$) to Lytle Shaw or Emilie Clark.

Shark seeks essays, reviews and art projects that take place in the journal. Our issues are organized thematically. Issue #5 is on collaboration. We encourage prospective contributors to contact us in advance. We do not publish poetry. Please direct all correspondence, including subscriptions and back issues, to:

Shark

> 74 Varick St. #203
> New York, NY 10013
> E-mail: shark@sharkbooks.com
> www.sharkbooks.com

Shark is also distributed by:
> Small Press Distribution, Inc.
> 1341 Seventh Street
> Berkeley, CA 84710
> E-mail: orders@spdbooks.org

ISBN:0-9664871-7-6

TABLE OF CONTENTS

Selections from *Index*

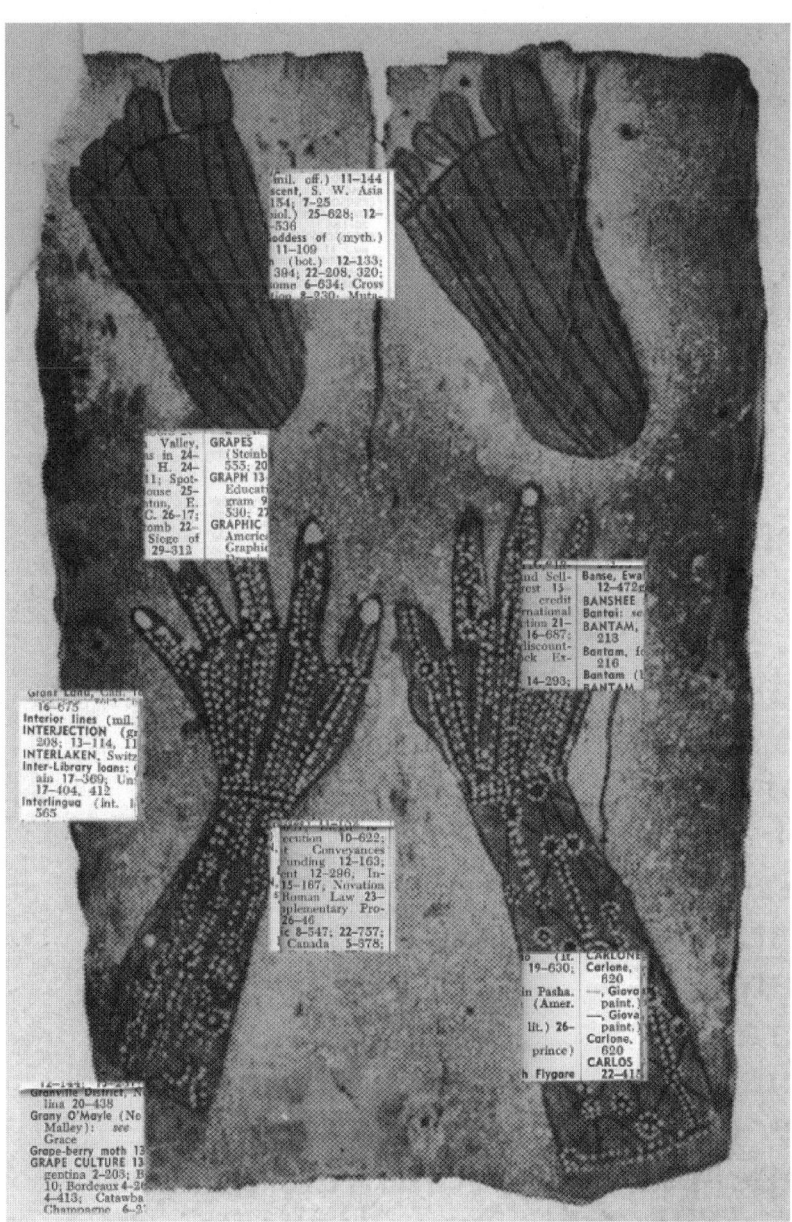

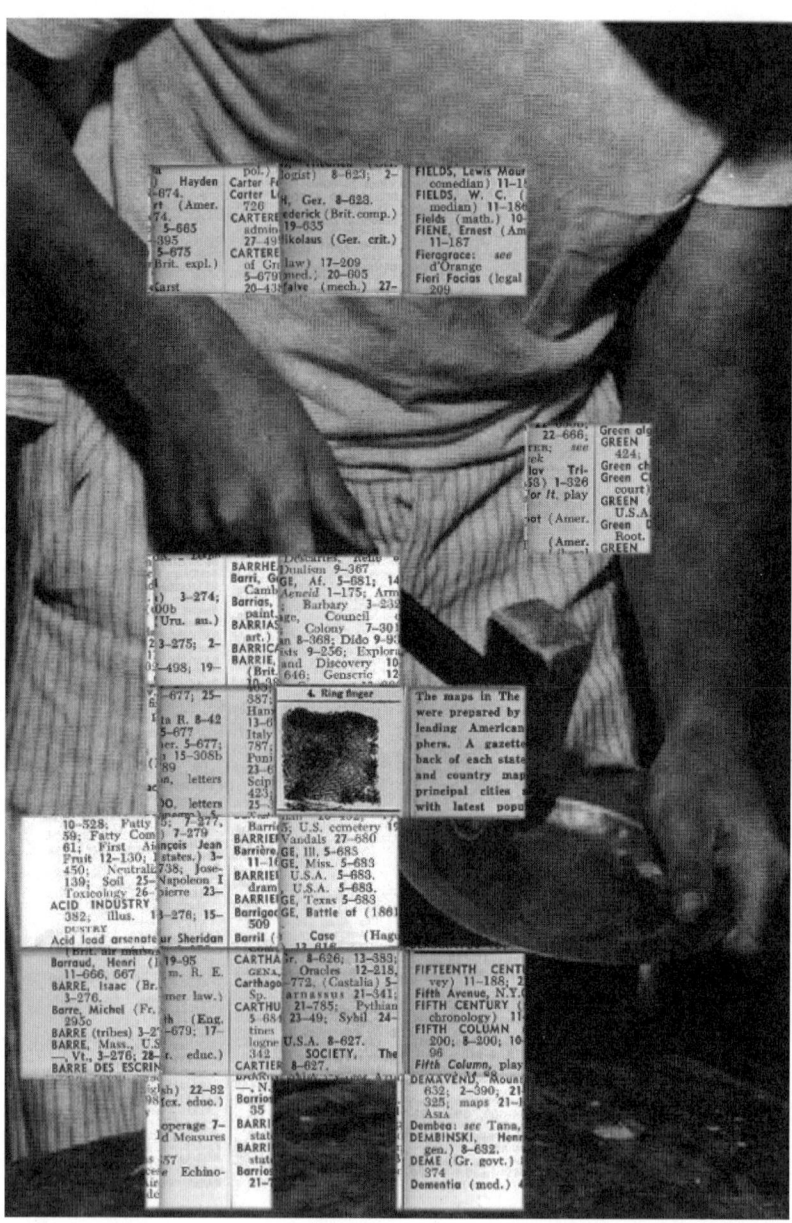

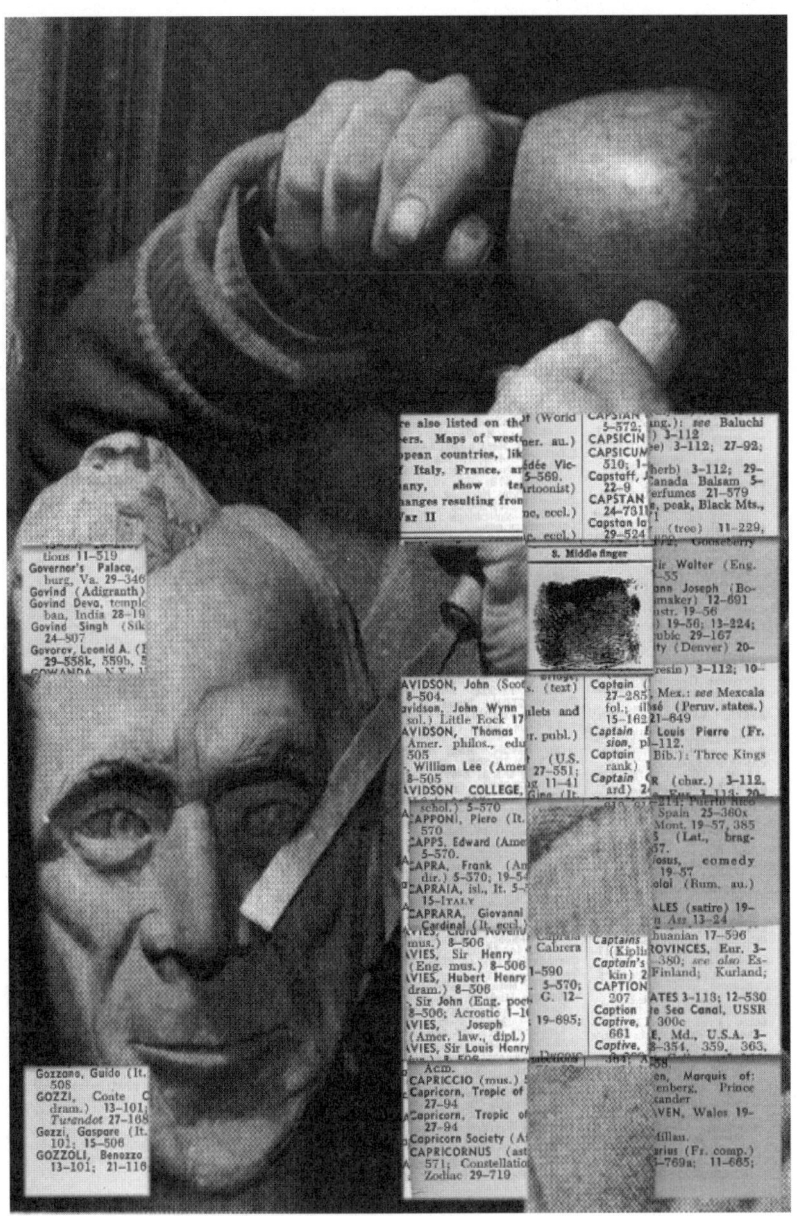

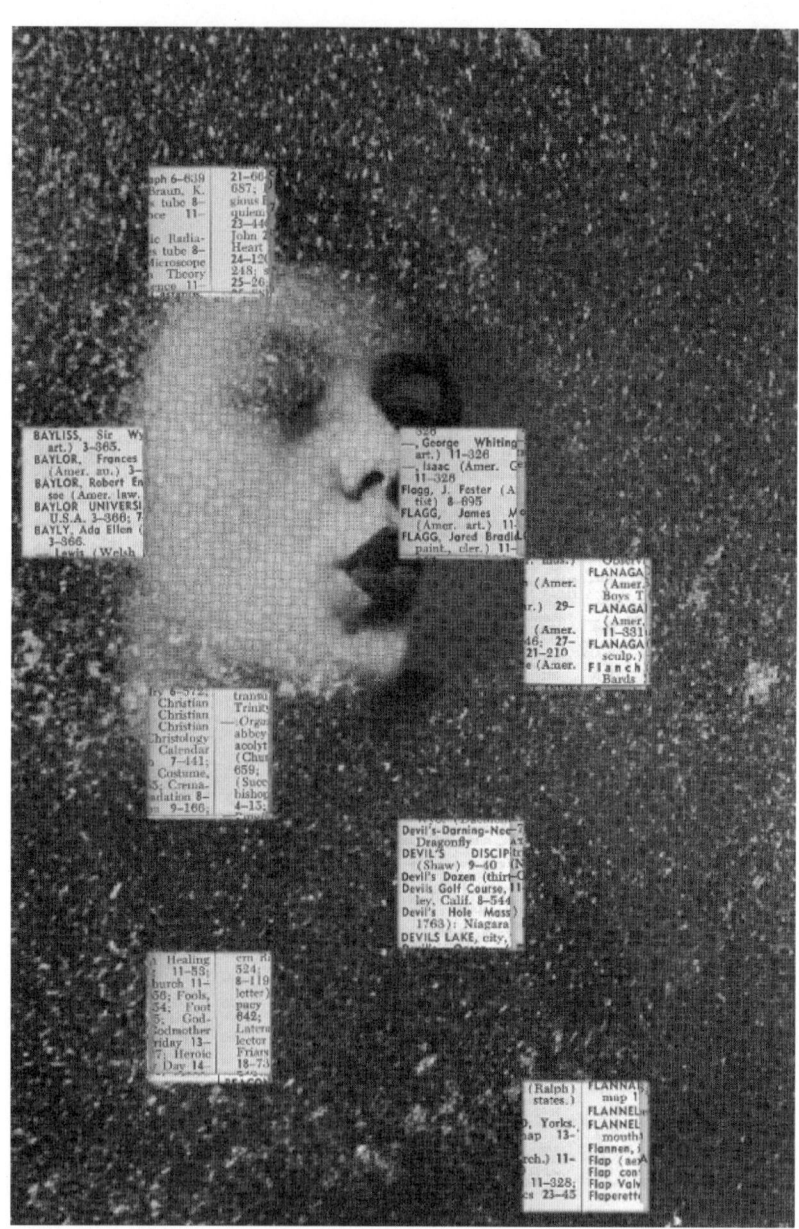

12 SHARK

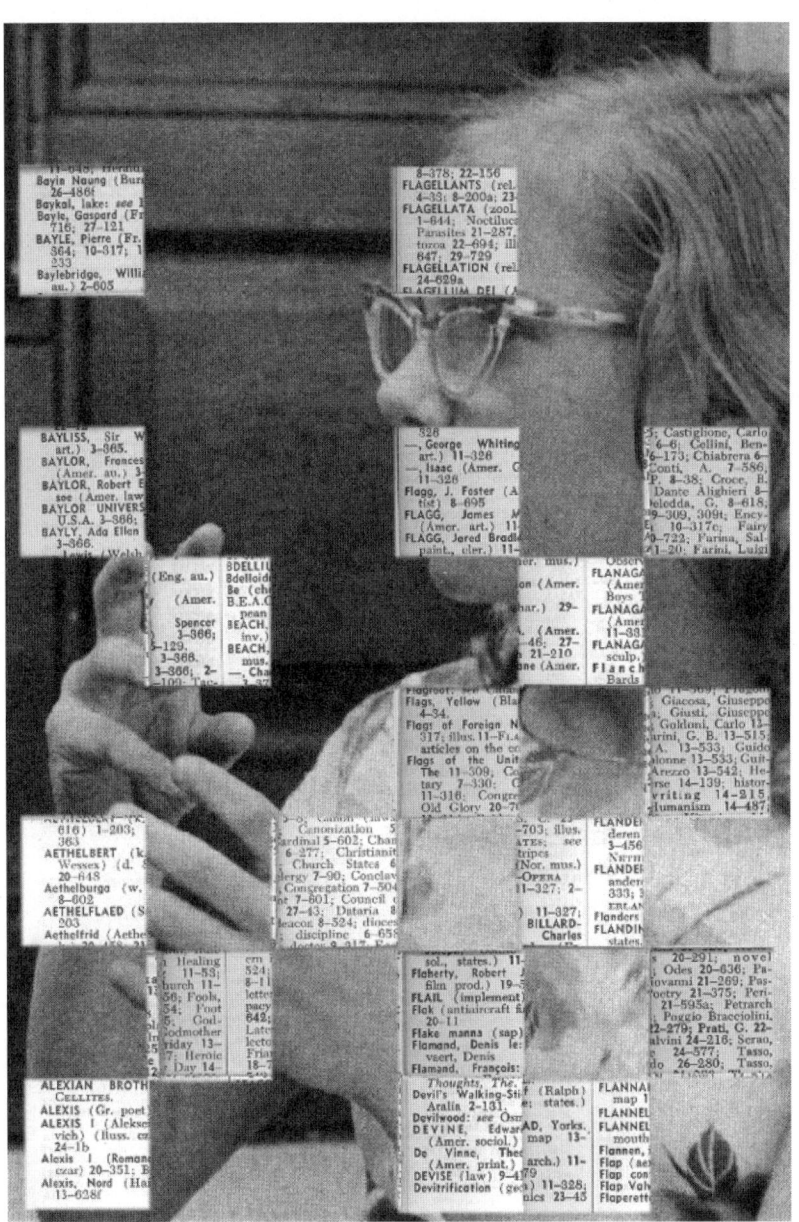

Scott McCarney 13

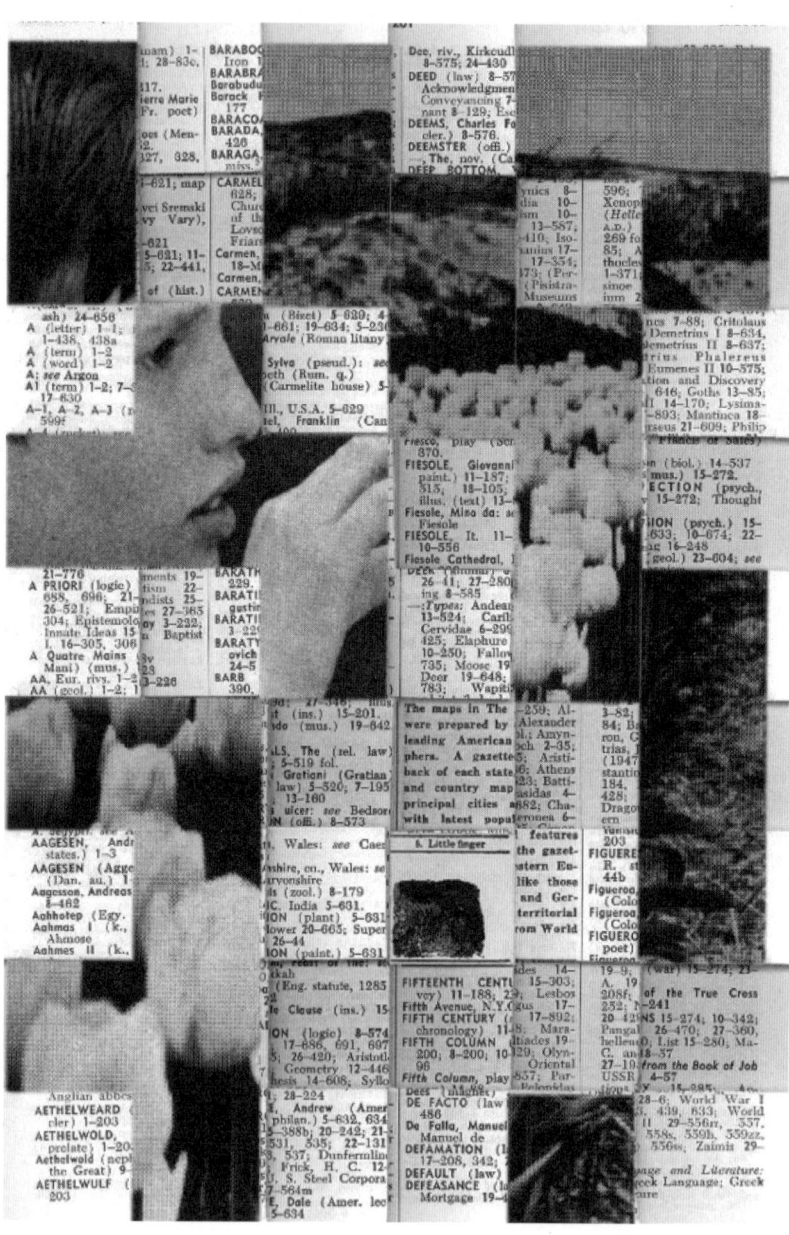

14 SHARK

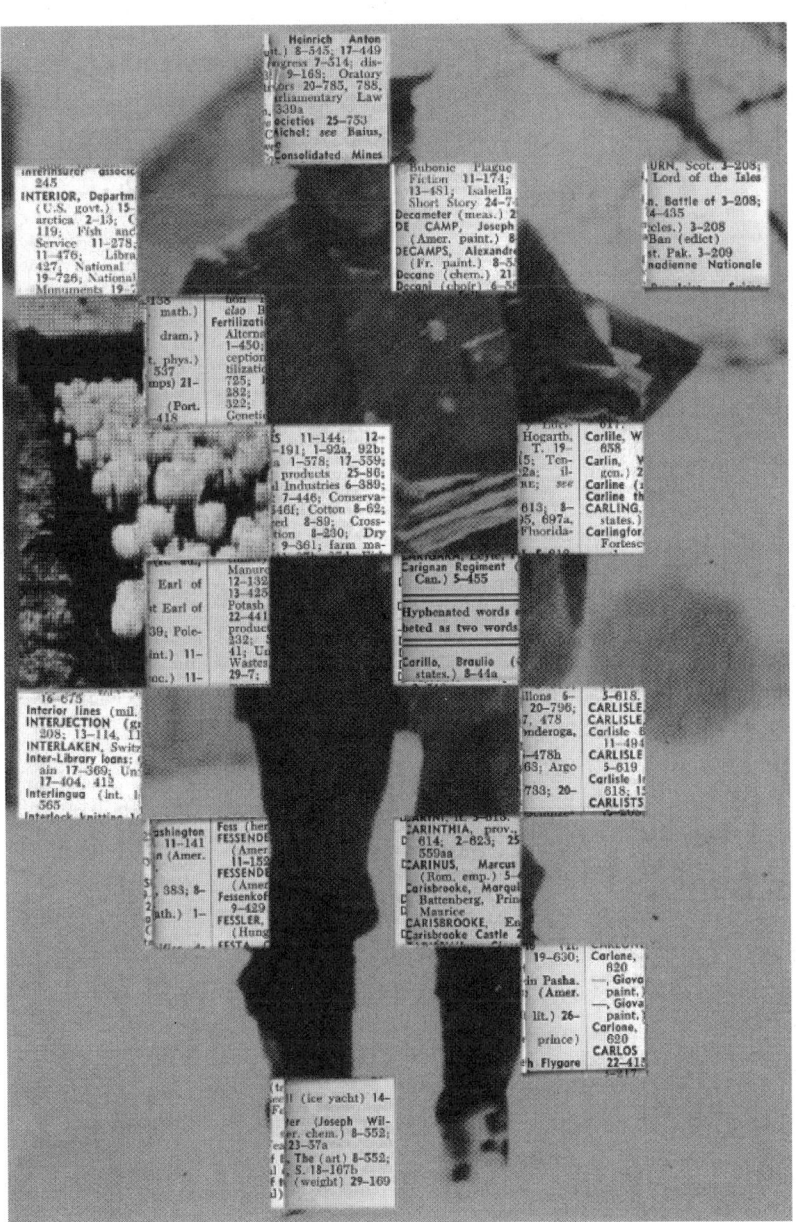

Scott McCarney 15

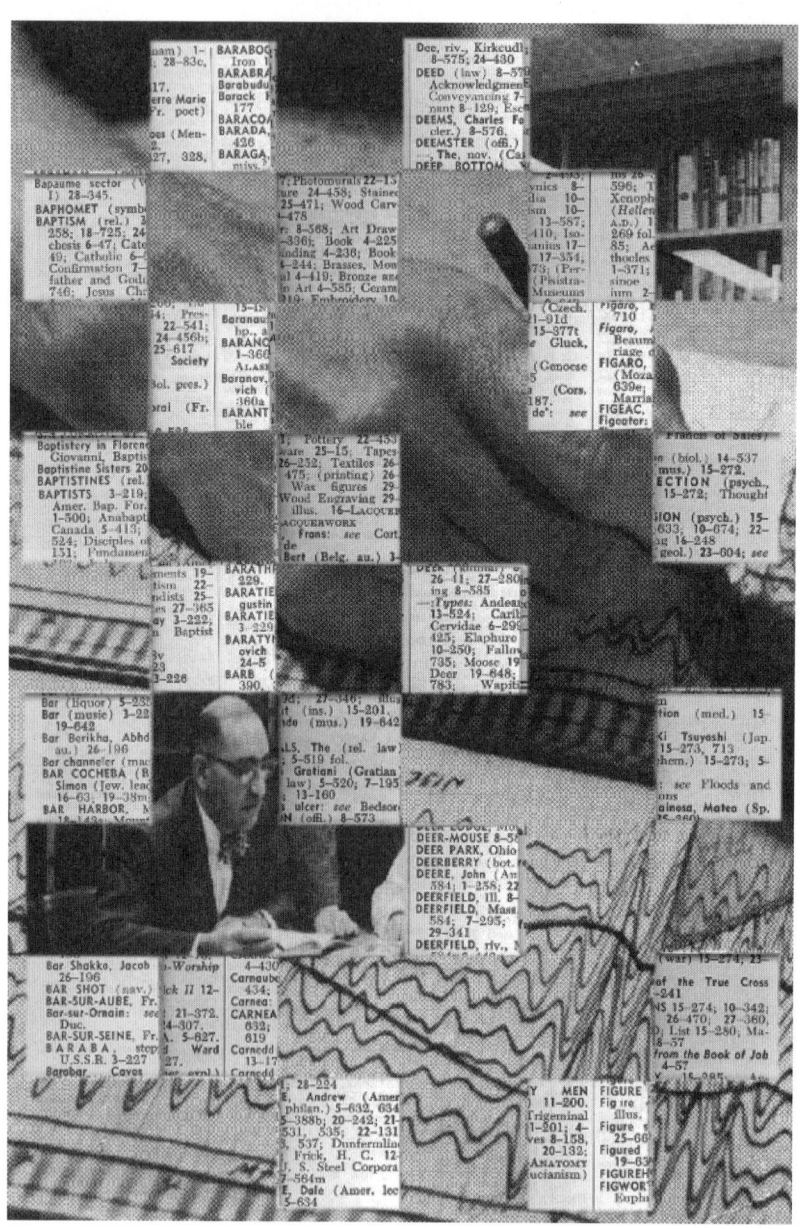

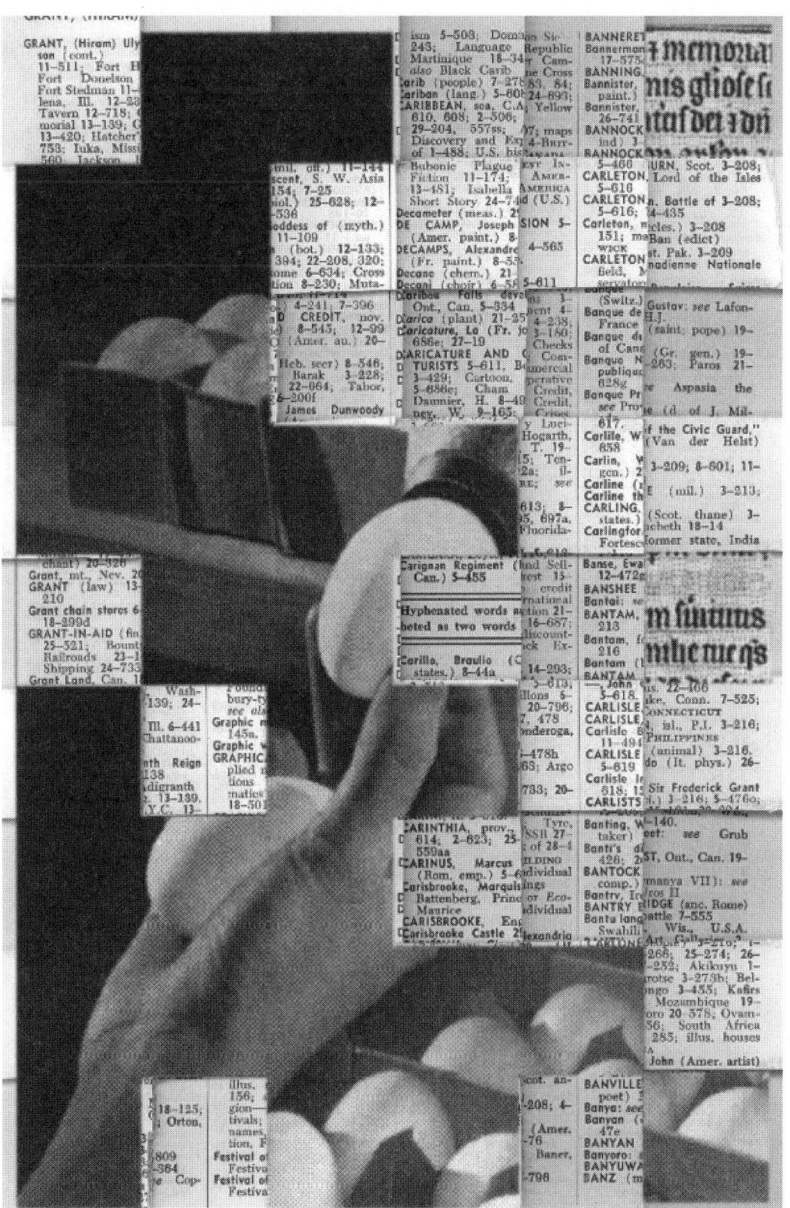

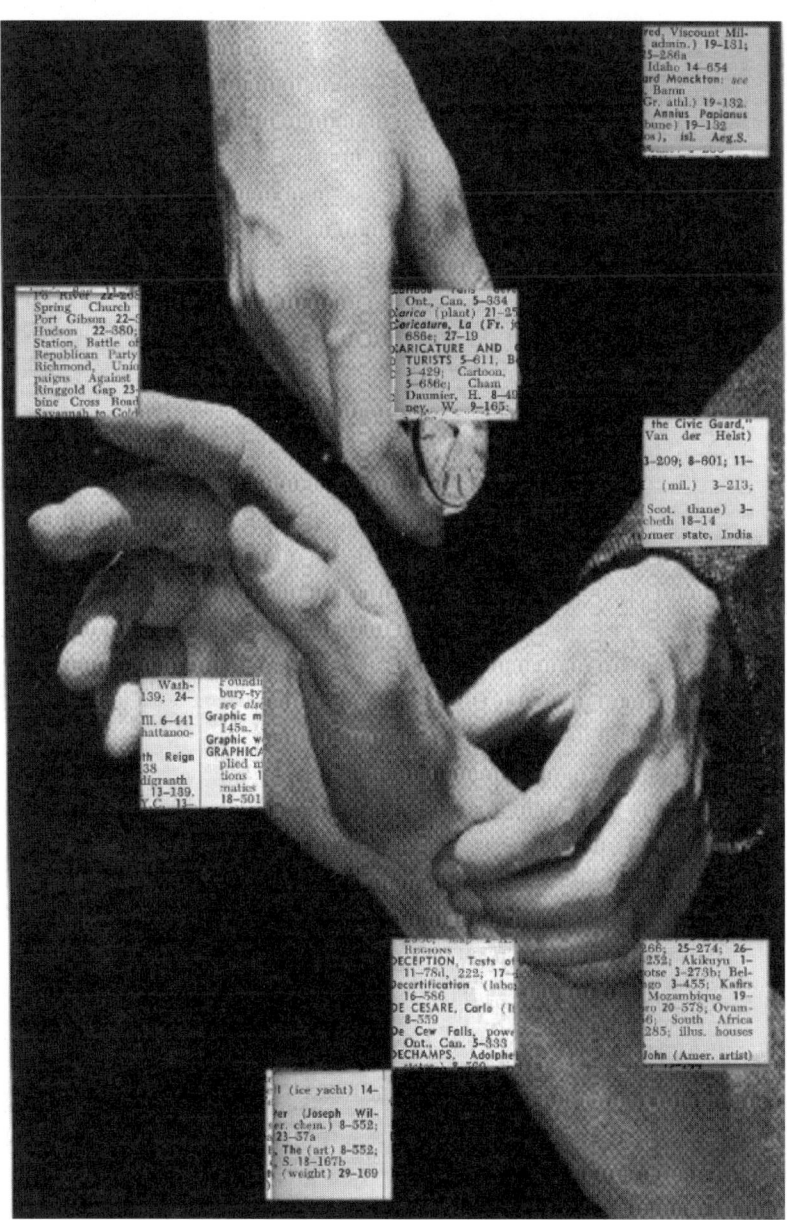

Words, Worlds and Procedural Form:
The Poetry of Inger Christensen and Klaus Høeck

For many people, reading translated poetry is like listening to music with your ears stuffed full of cotton. It doesn't make a lot of sense. What one gets is not the real thing, but rather a faded, distorted impression, just enough to make you feel the quality of the original, but never actually give you the full aesthetic experience of it. Whether or not one agrees with this view of translation, the problems and paradoxes of translating poetry are multiplied when it comes to poetry using procedural form. It's hard enough to reproduce metrical forms (or to convey the elegance of Rimbaud's "je est un autre"—"I is another"—into a language that does not inflect its verbs), but how do you even begin translating a book like George Perec's *La Disparation*, which consists of more than 200 pages of prose without the letter "e"? In such cases the virtual impossibility of translation becomes literal: either you break the rules, which more often than not form part of the work's *raison d'être*, or you forget semantics and create a new work, following the procedures, but not the content of the original.

The resistance to translation may be one of the reasons that—with the partial exception of the French-based OULIPO—the tradition of writing in procedural form has never become an international movement. This is so even though the use of formal procedures over the last thirty to forty years has become a widespread strategy in experimental poetry, and even though contemporary uses of procedural form are generally motivated by the same impulse: a desire to go beyond both a romantic-modernist conception of lyric subjectivity and its traditional counterpart, the imitation of existing literary models. No matter how general its concerns may be, however, by its very nature this type of poetry often ends up literally a prisoner of its own language. Which has had the sad consequence that experimental poets in one language-area rarely have a clue about the type of procedural work made in another.

That, of course, is regrettable. It is even more so since the uses of procedural form seem to differ significantly from one literary tradition to another. This means more than the fact that American, French and Danish experimental poets may use different kinds of formal procedures. Rather, it means that even though the use of procedures points to a general movement *away* from an aesthetic model of subjective creativity, it does not follow that all uses of such procedures point toward the same alternative. Developing out of French surrealism, the OULIPO, for instance, saw formal procedures as an upgraded version of automatic writing. They were considered tools, playful mechanisms with which one could probe hitherto uncharted areas of language: a fun way to explore semantic vistas that lie beyond the capacity of the human imagination, so to speak. By contrast, the most prominent experiments with formal procedures in contemporary Danish poetry have entered into a complex dialogue with German idealism. Their works orient themselves toward a world outside of language and use procedural form as a technique for getting beyond the Kantian distinction between "Welt an sich" and "Welt für uns"—between "World in itself" and "World for us."

That, at least, is the case for Klaus Høeck (b. 1938) and Inger Christensen (b. 1931), the only two living Danish poets who have consistently used formal procedures as a poetic strategy. Expressing a large-scale, inclusive aesthetic—Høeck's books are often more than 400 pages long—both poets use procedures in ways that organize vast, encyclopedic grids into which almost any kind of information can be put. In contrast with (parts of) the tradition of the American long poem, however, for Høeck and Christensen an aesthetics of inclusion does not point towards the inclusion of history in poetry—does not point toward a reworking of the epic. Rather, it gives room to myths, philosophy, natural sciences (especially geography, chemistry, botany and zoology), political discussions and everyday life. If anything, they are latecomers to a classical, cosmological genre often exemplified by Lucretius's *On the Nature of Things* or Guillaume de Salluste Du Bartas' *Le semaine ou la création du monde* (The seven days or the creation of the world).

The following essay will focus on how Christensen's and Høeck's specific uses of procedural form relate to this inclusive aesthetics. A couple of preliminary remarks are necessary, however. First of all, even though both Høeck and Christensen are considered among the most important poets in Denmark right now, neither of them has had much of an impact on younger generations of Danish poets. Neither are they—or the use of procedural form in general—considered a "school" in Danish poetry. In fact, in spite of significant similarities, the two poets don't even seem very interested in each other's work. Secondly, the few actual translations of Christensen and Høeck into English have consistently focused on semantic rather than procedural aspects. This not only means that the formal specificity of their poetry has been effaced, but also that the most experimental parts of their work are unavailable to people who do not read Danish. For pragmatic reasons, I have chosen to focus on works that are at least translated into English, if not generally available in the United States.

The Rhythm of Deep Ecology:
Inger Christensen's *det* and *Alphabet*

Inger Christensen published her first book *Lys* (Light), a slim collection of short, lyrical poems, in 1961. Her first major publication—which was also her first use of formal procedures—came out in 1969. Its title, *det*, has several possible meanings. Depending on the context, the word can be translated as "it," "that" or "the," that is: understood as a definite article, an ungendered personal pronoun or a demonstrative pronoun. "Det" is often used as the grammatical subject in phrase constructions that refer to continuing processes or states of being which have no agent or causal subject: the Danish sentence "det regner" means "it rains."

Referring to an outside world, but doing it with an almost immaculate lack of specificity—and that with a word which for lack of better is used to represent phenomena that do not fit into a logic of subject and verb—the small, unspecific title of the book points to its major concerns, since *det* is best described as an enactment of a cosmology: the creation of a world. To be more precise, the poems of *det* criticize structures of bureaucracy as they appear in everyday life and language and conversely bring forth an alternative that has everything to do with conceiving the world as a continuing process.

Criticism of the internal power relations of language was (and is) not a novel way to mix experimental poetry with an engagement in broader cultural concerns. What makes *det* unique is its focus on biology as a utopian resource. In *det* biology is understood in the broadest possible terms and described in a detached, almost scientific vocabulary. The contrast to rectified culture is not as much the creativity of the mind or the intensity of a life lived to the extremes, but rather a flowing, yet ordered "process" of the world as it is. In a later essay, Christensen has formulated it this way: "The forms exist … in the world. A tree exists in its tree-shape, and therefore my life, or the life of my family, can take on the same shape. But not in the way of a comparison, rather as a form that is the same. And which also could be a form of poetry. And here forms shouldn't be considered as something static, but as continuing processes that once in a while are elucidated." One could call this process "life," but it should be noted that life here is seen from a decidedly nonhuman perspective. Humanity— imagination, decency, language, madness—is for Christensen nothing but a small fragment of a larger process, and a fragment that ought to mirror the encompassing principles of natural creation. Culture and language should not be conceived as something apart from or as a contrast to nature, but rather as an extension of it.

As a consequence, political, biological and linguistic thematics overlap in *det*, often acting as metaphorical counterparts for each other. Or sometimes language and biology simply blend into each other, making natural procreations of words and/or a nature that moves with a laboratory-like artificiality. As in the poem, "THE SCENE, transitivities, I":

A word flies up tentatively flocks follow
at random massive biological forma-
tions. As if all this is really about reassurance All this
is really about an outer limit   desolate/ not desolate

a word that flies up   flocks that follow
neither more or less   birds that
fill this endlessly disappearing space
with a lack of explanation

All this is really a very vague explanation Do maintain that explanation
Turn on the wind-machine and let angels with
flapping wings be moved supreme as satellites

Let fleets of strangely dull creatures be
swept with the wind   insects with sails big
and torn as shiny illusions stand as
a vision: being's resistance to purity
        (my own translation)

Notice how words and phrasal parts are repeated and elaborated as the poem goes along, an elaboration that is both formal and semantic. What starts out "tentatively" somewhere perhaps desolate ends up as a whole flock of different types of floating creatures. This movement runs parallel to a gradual fictionalization, however: as part of the phrasal and semantic evolution the poem moves from a generic, unspecified space described in simple sentences and into a full-blown but artificial world that is also an "explanation"—perhaps a world-view—which needs maintenance. One should read this structure of gradual elaboration as an aesthetic enactment of biological concepts of evolution, mutation and deterioration. And there is no doubt about the significance of these concepts for *det*—and for Christensen's mode of writing in general.

Many of the poems and parts of *det* are filled with such 'formal narrations.' These are often controlled by procedural form. As if to exemplify "being's resistance to purity," the smaller 'narrations' do not add up to an overarching, global structure; neither do their procedural counterparts. Even though the structural architecture of *det* is quite elaborate, it works more like a static grid, often influencing the themes of a given poem, sometimes orchestrating faint echo-effects between different parts, but never in itself conveying a feeling of dynamic "process." Local as they are, however, the formal procedures of *det* have a double function. Being a form-giving logic that works independently of normal semantic considerations, they criticize—through distortion—the inherent power relations of everyday language. At the same time they enact a complex "process" that is given to the reader as s/he literally follows the rhythms of repetition, elaboration and deterioration in the parts and poems of *det*.

The same can be said about the formal procedures in Inger Christensen's next major work, *Alfabet* (1981, published in the United States as *Alphabet* in 2000). Only 80 pages long, *Alfabet* is physically smaller than the earlier book. It is also far more focused, though it shares both the critical and the cosmological ambitions of *det*. The most precise description of *Alfabet* is that from a simple point of origin—the first line/poem goes "abrikostræerne findes, abrikostræerne findes" ("apricot trees exist, apricot trees exist")—it charts the simultaneous unfolding and undoing of the existing world. All this is done in a poetic form that explicitly sets itself as an alternative to a paralyzed language and a self-destructive civilization. (*Alfabet* is very much written in the shadow of cold-war nuclear *angst*). This form is also one that tries to convey the dynamic process by which the world comes into being rather than impress the reader with a static conception of how the world is structured. The intricate web of formal procedures of *Alfabet* is central to that undertaking. From the outset conceived—and experienced—as a growing, roughly alphabetical list of things that "exist," the number of lines in each part of the book is determined by the Fibonacci series (1, 2, 3, 5, 8 and so on), while the things named are determined by the alphabet. Thus, the first part of the book has one line, naming existing phenomena starting with a. The second part has two lines, naming phenomena starting with b and so on. This alphabetical determinant, however, is almost effaced by translation—hydrogen, for instance, starts with b in Danish ("brint") and is therefore mentioned in the second part of the poem.

In the course of the book the original determinants begin a gradual process of both elaboration and deterioration. For instance, the seventh part of *Alfabet* (g) is substructured into stanzas that have the length of 1, 2, 2, 3, 3, 5 and 5 lines, thereby both repeating and distorting the Fibonacci series. This substructuring principle continues into the rest of the book, the eighth part—h—having the stanzaic structure of 2, 3, 3, 5, 5, 8 and 8 lines. Complicated by procedural echoing structures, a similar process goes on both in the syntax and in the semantics of the book, starting already in the second part, but becoming visible especially from the sixth (f) and on. Again, some of the 'formal narrations' are effaced by translation, but note how the first line of the book is inserted and repeated in this later poem, an insertion determined by one of the book's procedures:

> errors exist, instrumental, systematic,
> random; remote control exists, and birds;
> and fruit trees exist, fruit there in the orchard where
> apricot trees exist, apricot trees exist
> in countries whose warmth will call forth the exact
> color of apricots in the flesh

Already elaborated to include descriptive additions to the listed phenomena, from this point on both the list form and the alphabetic determinant gradually deteriorate and/or are elaborated upon. One could perhaps say that they evolve into other, less visible, determining principles, an evolution that adheres to its own immanent logic.

These largely formal movements are central to the experience of reading *Alfabet*—especially because, beginning with the name of a plant, the book gradually (part for part, sometimes stanza for stanza) includes new areas of being into its list of existing things. The total effect is that rhythm, syntax and semantics cooperate in order to convey the book's unfolding of a world. An unfolding not only read, but also felt by the reader who would thereby become one with the pulse of the world (at least this is the ambition) while reading its creation in the words of Christensen's book. *Alfabet* thus ideally lets its reader become part of a conception of cosmological time that is neither linear nor circular, but rather a kind of recursive and elaborate out-spreading of a distinctive pattern—much like that of the bracken, a plant that plays a major role in the imagery of *Alfabet*.

The procedural forms of *Alfabet* are the book's attempt at the rhythm of something like deep ecology. This, however, also complicates the overall cosmological and critical ambition of Christensen's work. As already mentioned, *Alfabet* runs from a meager 1-line beginning to a dizzying coming-to-being of the world, to a split between the destructive powers of a paralyzed civilization with its weapons of mass destruction and the continued "process" of poetical language. In this process both destruction and creation are conceived as phenomena 'born' out of the overall cosmology of the poem. Nuclear bombs and defoliants are not just man-made elements, alien to the natural world, but rather—as a consequence of Christensen's radical conflation of nature and culture—both a break in the rhythmic creation of the world and the logical extension of that same rhythm. This is also the case with the lyrical subject (which explicitly shows up in the tenth part of the book, but has been hinted at long before) and poetry in general.

The end result is not as much a denial of human responsibility as a bleak ambiguity, clearly expressed in the fact that *Alfabet* stops *in medias res*, in the middle of its fourteenth part. And in the fact that the book's last couple of poems do not see themselves as a trustworthy alternative to the ever more imminent threats of destruction. Apparently it has also been a paralyzing ambiguity for Christensen herself. While unanimously considered the most important living Danish poet, and selling many thousands of copies of her books, she has more or less stopped publishing poetry (only one chapbook and a couple of single poems in magazines since 1981) and has repeatedly stressed the fictional elements of her work, calling the books experimental as-if universes with no real importance.

Non-Human Epiphany: Klaus Høeck's *Hjem*

If Inger Christensen's oeuvre is small and condensed, then Klaus Høeck's is the exact opposite: overwhelmingly large, spreading out over more than 20 books, some of which easily exceed 400 pages and most of which are heavily influenced by his own variant of procedural form. In Høeck's case, this use of procedures was originally inspired by cybernetic information theory and the proto-structuralist Danish linguist Viggo Brøndal: having read that the style of any given poet was statistically quantifiable if one charted the frequency of different word groups (nouns, verbs, adjectives and so on) in his or her poems, Høeck tested the thesis on his own early work, found it true and set out to create formal procedures that would either force him away from or exhaust his own style, determined through statistical calibrations. Often conceived as variations upon a 'master text,' the result is works influenced by formal determinants of a staggering, opaque complexity.

The following examples are typical, but in no way exhaustive: taking a text of his own as a starting point, Høeck has often statistically determined its alphabetical and grammatical distance from the mean average of the Danish language (that is, frequency of word groups, dependent clauses etc.) and from that information written a series of poems, each of which has the same distance in one or more of the measured categories. Or he has written poems that copy the grammatical structures of each other, sequentially and/or numerically, or used a master text to create a probability matrix with which he, word for word, determines the succession of word-groups in the phrases of the next many poems. All this combined with other, more or less visible formal constraints, such as a consistent use of (sometimes) self-invented syllabic metrical patterns.

The complexity of all this could give the impression that Høeck's work closes in upon itself or becomes as dry as a high school math-book. That is not the case. On the contrary, all of his books are very explicitly oriented towards something beyond themselves, whether that something is limited to other works of literature and art (in his early works), extended to politics (in his works of the late seventies) or—as in the eighties and nineties—defined as an outer world in its non-human totality. At the same time, his work revels in different stylistic modalities, ranging from total gibberish to philosophical anecdotes, landscape description, warm, often self-ironic humor, strange imagery and high pathos. All of which points to a legacy both from English and German romanticism and from high modernism, a legacy found in large parts of Danish counter-culture. In that sense, Høeck does not reject an earlier literary tradition, but reworks it in a strange, pseudo-scientific setting. In his own words, his poetry creates "a tangible rose full of early snow on transcendental graph paper."

Beginning in the late sixties, Høeck's publications over the next ten years include some of the strangest, if not the best, books ever written in Denmark, including *Project Perseus* (1977), a collection of "sci-fi poems" partly written in the computer programming language Algot 8 and *Ulrike Meinhof* (1978), a 4-part double wreath of sonnets that is also a eulogy to the German terrorist group Rote Arme Fraktion. His *opus magnum*, however, is generally reckoned to be the 600-page, monumental *Hjem* (Home), published in 1985.

Hjem is a book about Denmark, or rather a poetic-encyclopedic glossing of every thinkable aspect of Denmark conceived as a self-governing system: among other things there are sections on meteorology, geography, zoology, architecture, the industrial production, the management of waste and, but only as a part, the human inhabitants. All this in poems that are visibly marked by procedural form. Consider this poem describing Samsø, a Danish island which is popular as a summer resort. Notice the rudimentary, involuted syntax, a clear sign of one of Høeck's formal procedures:

> Samsø. If one. Camping.
> Now cooks Kosan-gas. Before: child
>   hood. Erosion of language mor
> aine. Ah, paradise with dune
>   and children's slide. Pine forest's
> spray. Between an attic
>   where God lived perhaps.
> Or distant thunder in lathyrus
>   bursts lightning soon. In your
> memory: squid castle.
>   (Translated by John Irons)

This poem highlights important stylistic and thematic differences from Christensen's work. First of all, procedural form does not produce any kind of rhythmic pulse here: in Høeck's poetry metrics and syntax are orchestrated in ways that suppress any connection to a rhythm of a living body. Instead, the distortion of grammar and the syllabic metrics create a poetry whose lines and sentences act as strings of empty slots, eventually filled out by the joint effort of Høeck and his procedures. So that any poem is a matrix rather than an organic form, human or non-human as that may be. Secondly, even though the overall ambition of *Hjem* is very much related to Christensen's poetic intentions, the "world" that the poems relate to is far more particularized. In fact, it is less the "world" so much as very specific instances of it, fixed in both time and space, that is the theme of the many poems of *Hjem*. More generally, and in his own words, Høeck's poetry is extremely "lococentric." One could translate this phrase as site-specific.

In more general terms *Hjem* (like most of Høeck's later books) *does* share some concerns with Inger Christensen's work. This becomes evident as soon as one looks at the global architecture and layout of the book. *Hjem* is split in three parts, or rather: three strings of poems, each with its own procedural structures, and each named after its thematics, namely "Nature," "Culture" and "Spirit." The strings are not isolated from each other, but run in the top third (nature), the lower third (culture) and the middle of the book (spirit), so that every page of *Hjem* holds three poems. The engendering master-text for "Nature" is the different formulas for all minerals naturally existing in Danish geology. "Culture," on the other hand, is born out of a nonsensical letter-string, which statistically reflects the frequency of the different letters in the Danish language. The middle string, "Spirit," then, "takes" its words from the two other poems on its page, thereby trying to enact a sort of mixing, a mediation of the two realms—unfortunately, the process is somewhat effaced by any translation.

To give an idea of how all this ends up, one might take these three poems, which come from the same page (324). As a testament to the difficulties of translating procedural works, one could point out that the Danish word for tooth ("tand") in its plural form is similar to the verb "ignite" in its present tense ("tænder"). Here is John Irons's translation:

> What's a protozoon? – No you've no
> idea. Can it be an epidemic disease
> of the lungs or of the skin between t
> he fingers. Perhaps a certain make of
> toothpaste? – In fauna and biology bo
> oks you'll find the protozoon.

> you look through this verse
> causing it to ignite at
> a certain instant

> It is a proud moment
> when you can drive an MZ
> locomotive through the poetry.
> When you hear the turbine

> turning in the metaphor,
>
> when you see the electric engine
>
> spark in the syntax,
>
> you know it. Almost
>
> 3000 hp put into
>
> verse. It is a proud moment.

As one can see, each of the three strings has its own thematic specificity, letting microfauna, locomotives and metapoetic statements meet on a single page. One should also note, however, that formally the middle string in *Hjem* works more as a conceptual nod towards Høeck's implicit aesthetics than as the actual site of the meetings between nature and culture in the zone of spirit and/or poetry. Instead, such epiphanic meetings are spread out all over all three strings, as sudden glimpses acted out by strange metaphors, procedural distortion or high-strung addresses to the reader. Which also means that even though Høeck's epiphanic ambitions relate directly to romanticism and to a certain line of modernist poetry, he does not enact these ambitions through recourse to symbolic language or a poetry of a phenomenological perceptiveness directed at the outer world. Through its procedural form *Hjem* creates a new sort of epiphany, whose center is not so much the human mind, but a conflation of structural or ontological zones (nature and culture) set in motion by the procedures and, if he has anything to do with it at all, only fleshed out by the creative powers of the poet. In that sense, *Hjem* is a site for non-human, but epiphanic, moments.

Høeck's later poetry has in many ways followed in the tracks of *Hjem*. It has, however, become less jagged and disjunctive. Not that the procedural forms are no longer there, but they have become assimilated into the modalities of a poetical subject speaking in many tongues—or, as Høeck himself states it in a poem from the book *1001 digte* (1995, 1001 poems), they have become diluted into his poetry like salt in sea water. All this has had the strange consequence that Høeck, whose poetic career started out with a criticism of individual style, now has an easily recognized stylistic signature. Connected to this, but also to the ambition of a "lococentric" poetry, a less paradoxical, but still surprising development is the fact that Høeck's latest books are unabashedly autobiographical, his latest, *In nomine* (2001) simply being a poetic autobiography using his full name (Klaus Høeck Johnsen) as the engendering master text. Still, neither the lyrical subject nor its language stands as poetry's ultimate point of origin. Instead, the biographical "I" becomes part of the site of the poem—"Klaus Høeck" reemerges as a figure in his poetry, but not as its creative starting point. One could say that his books investigate how their procedures, forms and themes collide with the time and space of their making—and that the life of Klaus Høeck is a natural part of that time and place. Which gives a sort of biographical license to this late oeuvre, without letting it fall into the subjective pathos of an earlier lyrical tradition.

Note

For references to translations of and other information about Inger Christensen and Klaus Høeck, see www.litteraturnet.dk, which exists also in an English version. Høeck has also published occasional and political poetry on the internet: http://kurdistan.life.nu and http://imagine.stop.to.

Matthew Buckingham

Twogether Beside Himself: *More Details From An Imaginary Universe*

Most people consciously and unconsciously construct personal, private systems for understanding the world and navigating life. Few people deliberately reflect on these internal idiosyncratic 'languages,' and fewer still make them publicly known. Matt Mullican is someone who has. Over the last thirty years, in the form of his art work, Mullican has disclosed the details of a personal cosmology in sites ranging from museums and galleries to subway stations and public parks. This has not taken the form of a temporally-bound autobiography, but rather a semiotics of subjectivity—traffic-lights and signs at the intersection of perception and identity.

Historically, the work of Mullican, and other artists of his generation who were also concerned with questions of perception, can be seen to form a bridge between the phenomenologically-based work of the previous generation (particularly the minimalists), and a new interest in perception as it related to representation. For many of these artists this shift critiqued and positioned phenomenology in relation to the social and political, leading to direct, explicit examinations of identity. Mullican developed a typology for life-experience which he continues to express within a language of signs. These signs deploy a wide range of media, each correlating specifically to the cognitive order he asks them to describe: graphic banners and posters represent *symbolic* systems structured on language; techniques of drawing are

WHAT IS THIS MAN THINKING?   Matt Mullican, *Untitled.*

Matt Mullican, *Untitled*, 1982.

used to describe *imaginary* visual and spatial experience; and photography and video delineate indexical traces of the *real*, which is elusive—always mediated—a residue outside of language. Mullican's quasi-hermetic semiotic discourse produces a dialectic of signs, neither truly personal nor truly public, which, in turn, suggests a triangulation between Mullican (as a 'self,' subject), the world at large, and the signs that are produced by each. By submitting his personal, subjective reality to self-conscious examination, Mullican, in effect, stands beside himself, deliberately doubling his perceptions by expressing them externally, symbolically. His art work catalogues these signs which, in turn, archive his experience. And even though the project is encyclopedic in scope it's ultimately more flexible, more accommodating, because it does not merely record experience, but instead documents mental processes—specifically the system he employs to create his cosmology. Deciphering 'Mullican-World,' we begin linking it to our own, suspecting that our own cosmologies might also be as easily incorporated into his, or anyone else's. In trying to *know* reality we find that we *are* reality. The title of Mullican's recent book, *More Details From An Imaginary Universe*,[1] provides an essential example of Mullican's methods. The title is borrowed from an art work of 1973 consisting of a series of ten irregular-shaped clippings from comic strips mounted on a white sheet of paper. Written dialogue and narration from the original comics has been excluded by Mullican's scissors. In the cut-outs we see the wheels of a car; a commercial sign beside a road with a car in the background; a modernist house in silhouette; a group of people walking away; a similar scene; a fragment of a telephone receiver; a woman's face; two sets of breakfast dishes with eggs, bacon and coffee on a table-top between two pairs of hands; a woman pausing on

Matt Mullican, *Untitled (Matt Mullican under hypnosis; looking for an essential quality in a room)*, 1996.

a busy street; and a bouquet of flowers gripped by a hand at the edge of a door or window. The group of images presents a problem; what 'reality' are we experiencing when we contemplate the details of a world which *could* physically exist but does not?

A few pages away in the same book there is a reproduction of an untitled diagrammatic drawing, from two years later, labeled "outline of work" in which Mullican seems to be explaining his thought process to himself. The drawing itself is a sign for a system of thought and also a meditation on representation. In it he describes a range of possible ways to represent a person, a subject. He arranges these along a spectrum, a flow-chart of sorts, moving from right to left, from the concrete and indexical to the abstract and imaginary. First he contemplates the ability of a photograph to signify a subject, then he compares this with how a subject is represented by a drawing, followed by a stick figure, and then a codified sign (similar to those marking public toilets). Next he looks at the abstract shapes that comprise this codified sign, and so on. Mullican concludes his 'outline' with the most ephemeral but, at the same time, potent form of representation: thought itself, the purely mental representation of a subject as it exists within memory or dreams.

As with most taxonomies, Mullican's taxonomies-of-experience reveal more about their maker than about the world they claim to describe. This is the cleverness of 'Mullican-World.' In the hand-written explanations on the untitled drawing, "Outline of Work," there are a number of misspelled words (i.e., 'compleat,' 'definent,' 'eiether,' 'sences,' 'squair,' 'symblo,' 'twogether'), which raise an interesting set of questions in light of Mullican's concerns. These alternate spellings, some of them nearly puns, resonate with meaning. Were the words intentionally misspelled? If not, can they be read in relation to Mullican's project, or are they simple errors—slips of the pen? In any case these re-spellings point to ways that language (and the symbolic order) are always partly unfixed and malleable. If language fails to contain or express full and precise meanings, it's also true that when we speak and write we often fail language as well. The most complex and, at the same time, direct manifestation of this type of ambiguity in Mullican's work can be found in his hypnosis 'performances' and subsequent video documents of these events. Working with different hypnotists over the years in many locations Mullican has submitted himself to various hypnotic suggestions that have induced subconscious journeys to past, future and parallel realities. Hypnosis utilizes an unconscious state to access and generate narrative. The most interesting question, again, is "what reality are we seeing?" When we watch Mullican under hypnosis, what nested combination of the 'real,' 'imaginary' and 'symbolic' do we encounter?

The relation of narration to time is homologous with the relation that models and maps have to space. These forms condense, elide, and reorder perception to demonstrate a point. Both rely on a kind of fiction which is collectively taken to be factographic and true. People reformulate trauma, as well as joy, in the form of narrative in order to defuse, or amplify emotions. Similarly, children surrounded by a world of giants "...use play to create a world appropriate to their size. But the adult, who finds himself threatened by the real world and can find no escape, removes its sting by playing with its image in reduced form."[2]

In an interview between Mullican and Michael Tarantino, appearing in *More Details From An Imaginary Universe*, Mullican says that some people associate his work with toys: "...but what is a toy?" he asks. "A toy is a minimum amount of construction, because a child is supposed to imagine the rest."³ Indeed, toys and games are the "rhythms in which we first gain possession of ourselves"⁴ and they can easily be read as a codified, miniature, interactive encyclopedia of the adult world, which has been inserted into the child's realm of play. Toys are representations with social meanings which naturalize the larger cultural constructs they reference such as gender, war, and industry. Toys have also been called upon, from time to time, to help introduce new technologies to society, or at least to create an appetite for the 'new.' The first toy automobiles were manufactured in 1900—three years before Henry Ford founded the Ford Motor Company.

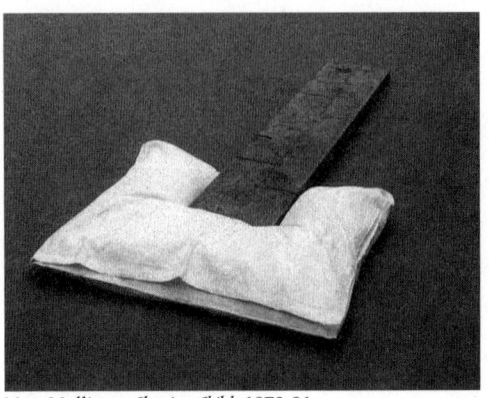

Matt Mullican, *Sleeping Child,* 1973-91.

But toys which are prescriptive may be valued more by adults than by children. This type of toy presents the conditions of the world as being inevitable and pre-existent, omitting agency and creating children who are users, not creators.⁵ This reveals a certain tension between the symbolic and the imaginary. As Mullican alludes above, for children, the most minimal materials are often the most flexible. Being less defined, these materials have the capacity to hold more meaning. To a child anything can be a toy or, literally, a 'plaything,' or in German a 'Spielzeug.' The gap between toys and the games children play with them is potentially enormous.

This ambiguity in the psychological economy of toys (latent in both representational toys given to children by adults, as well as everyday materials found and transformed by children) is illustrated very clearly in Harper Lee's Novel *To Kill a Mockingbird,*⁶ and perhaps even more so in the film adaptation of the book, where special visual emphasis is placed on the function of toys within the narrative.⁷ Much of the story revolves around a young girl named Scout, and her older brother, Jem, who are morbidly fascinated with their mysterious neighbor, Boo Radley. Jem and Scout have never seen Boo, a recluse who is reputed to have once stabbed his own father. Midway through the story Boo begins leaving small ephemeral objects such as a broken

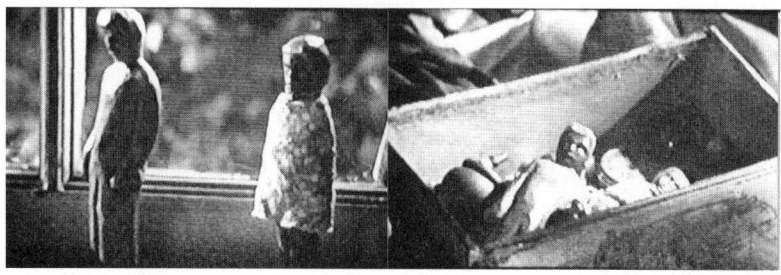

Frame enlargements from *To Kill a Mockingbird*, copyright 1962, Universal-International, directed by Robert Mulligan.

watch, dice, jacks, and a pocket knife, for Jem to find in the hollow of a tree. Jem avidly collects these items in a cigar box. This mute one-way system of communication culminates in Boo's gift of two hand-carved dolls he has made himself. The dolls are representational in ways that the previous treasure are not—specifically they represent Scout and Jem. For Boo the dolls are avatars of the children. To Jem and Scout they reveal that they are being watched by a stranger. The accurate details in the dolls' features make it clear that Boo is observing them very closely, but the dolls also reflect Boo's, and, in turn, the adult world's projections onto the children. The rendering of the clothes on the doll depicting Jem is fairly accurate, while the doll depicting Scout wears a dress, something Scout wears only if she is forced to do so. As signs for communicating between adults and children, toys are always potential sites of conflict and even resistance. As Susan Buck-Morss points out in *The Dialectics of Seeing*, for Walter Benjamin, childhood represented an unsevered connection between perception and action: "Children's cognition had revolutionary power [for Benjamin] because it was tactile, and hence tied to action, and because rather than accepting the given meaning of things, children got to know objects by laying hold of them and using them creatively, releasing from them new possibilities of meaning." Benjamin considered the adult repression of childhood and its cognitive modes to be a political problem.[8] By rejecting received ideas about representation and bringing to light individual and collective processes of perception, which are often kept hidden, Mullican ties perception once again to action, warning against mistaking symbols for ideas and reminding us of the dangers of unconditionally pledging our allegiance to signifiers.

Notes

1. Matt Mullican, *More Details From An Imaginary Universe* (Torino: Hopefulmonsters, 2000). For me the most evocative and complex text here is Allan McCollum's 1979 essay "Matt Mullican's World" (26-34) reprinted from *Real Life*, Winter, 1980.
2. Walter Benjamin, "Old Toys," trans. Rodney Livingstone, in *Walter Benjamin: Selected Writings, Volume 2, 1927-1934* (Cambridge and London: Harvard University Press, 1999), 100.
3. In the interview Mullican continues: "I think the toy relationship to my work is really interesting, no one ever talks about that, maybe because they feel it would be dismissive of the work." Far from trivial, the analogy of Mullican's work to toys is more than justified.
4. Walter Benjamin, "Toys and Play," trans. Rodney Livingstone in *Walter Benjamin: Selected Writings, Volume 2, 1927-1934* (Cambridge and London: Harvard

University Press, 1999), 120.

5. See Roland Barthes' essay, "Toys," in *Mythologies*, trans. Annette Lavers (New York: Hill and Wang, 1972).

6. Harper Lee, *To Kill a Mockingbird* (Philadelphia: Lippincott, 1960).

7. The 1964 film adaptation of *To Kill a Mockingbird* was directed by Robert Mulligan.

8. See Susan Buck-Morss' discussion of Benjamin's theory of children's cognition in chapter 8 of *The Dialectics of Seeing* (Cambridge: MIT Press, 1989).

Illustrations courtesy of Matt Mullican.   Special thanks to David Grey and Michael Tarantino who were crucial to the editing of Mullican's catalogue *More Details From An Imaginary Universe*.

Jovi Schnell

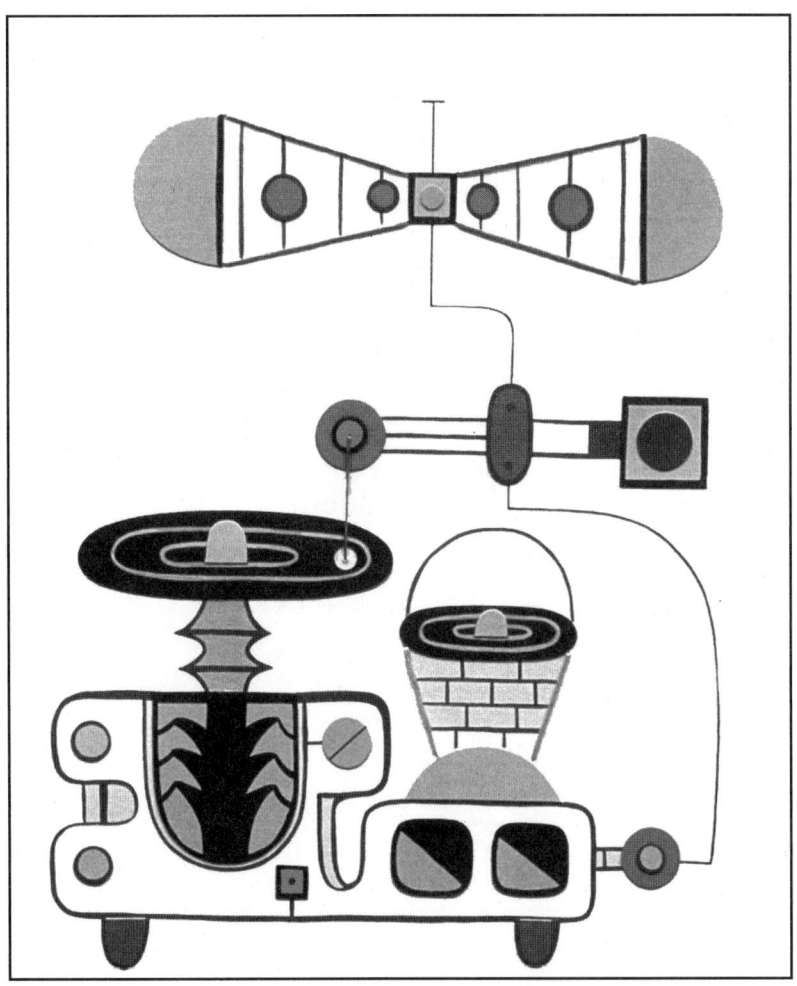

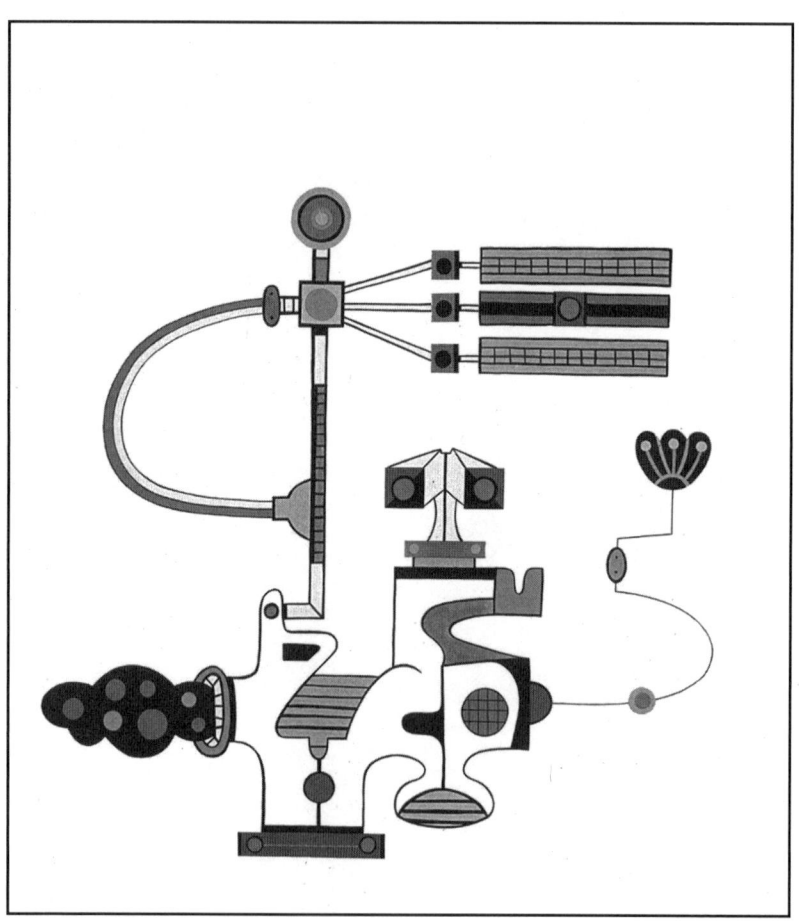

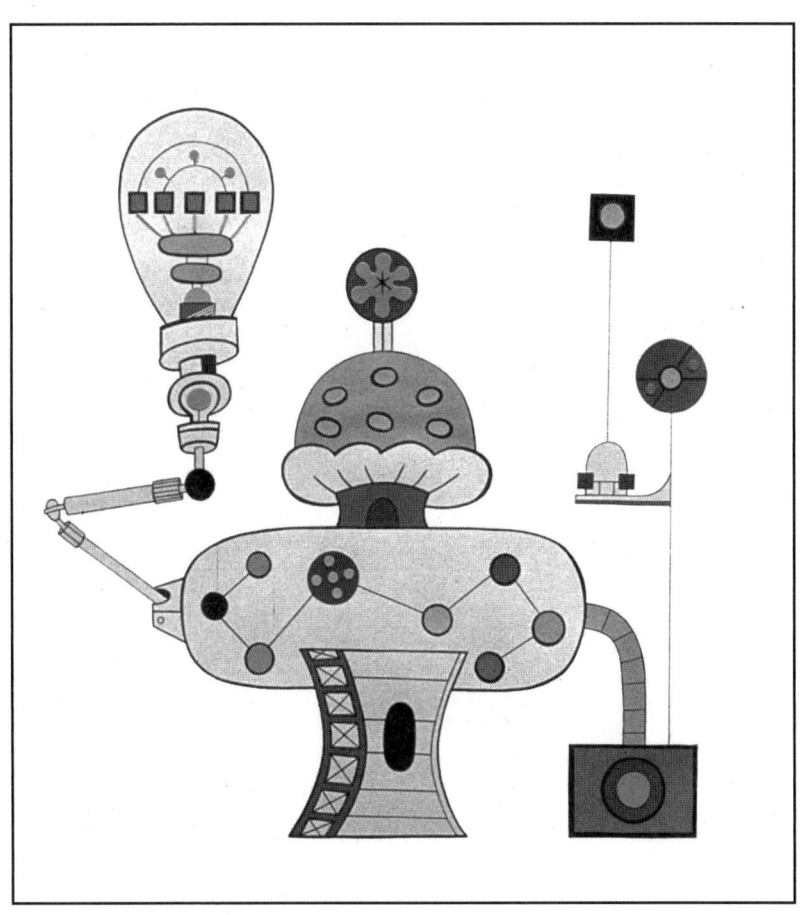

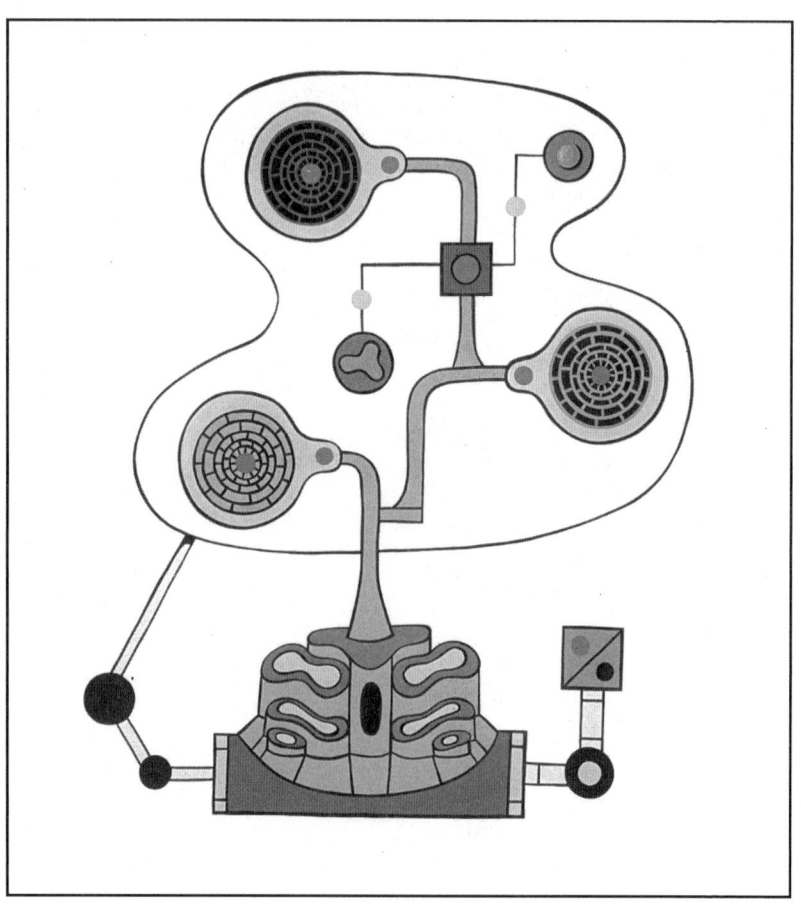

Titles:

Pg. 41—*Sweller,* 2000. Acrylic, thread, collage on paper. 9"x11"
Pg. 42—*Reporter,* 2000. Acrylic, thread, collage on paper. 19"x20.5"
Pg. 43—*Extension,* 2000. Acrylic, thread, collage on paper. 19"x20.5"
Pg. 44—*Formlet Freeze,* 2000. Acrylic, thread, collage on paper.
　　　19"x20.5"
Pg. 45—*B-Scape,* 2000. Acrylic, thread, collage on paper. 9"x11"

A Dumb Conundrum:
What I Like About Jason Rhoades &
Jason
Rhoades's *Volume A Rhoades Referenz*

A .

A.B.S. GUN WITH POM
FRITZ CHOKE AND
AQUA   NET
abstraction   the human capacity for
(IKEA CAR-
PETS)

as the grease slowly spreads autop-
urists

plugged into   an arti-
ficial

anus (ass-hole)

ALPINIM-
PALA
sonal) explanatory model for   (per-
the "redundant" quality of
american formalism
approach—you have to
drive a car right up to the building,
go in.   park there, and then

con/tent

deal   the way   we
with objects
from the past   and cultures   when we are

unable

to reconstruct their original

meaning

B.   BLUR
TEMPORARY ART   THE BMW SCHOOL OF CON-
Max Bier Baum
liked   cosmetics

liked artist's life cartoons
there are p r i n t a b l e
n u m b e r of im
BLUE ROOM AND LOVE SEAT

portant Do
ris D
ay
M Y B R O T H E
R /
B R A N C
USI
BUCKET LAMP
(TV, fish tanks, easy chairs)   Buckets
are never out
          of place.

C.

the root tuber of the carrot plant
to soup up, tune up, hot rod
chevrolet   sheet rock
Montgomery Ward Clinique Clinic
CHERRY MAKITA (HONEST ENGINE)
pin up girls constantly downloaded
simultaneously construction
and destruction, but
it is also about where this information
is stored. You can sit on
a massage chair and play a video
game,
a car radio, vehicles, tools
and assorted lamps.

D.

DESIGN ITALIA:
     Bean Bag Chair   Ferrari   espresso
machine
                         DIALOG
                              interplay between
          human beings   locales   objects
and
interference therewith
               Dictionaries are present—   and, to
          varying degrees, visible—

          in almost all of the artist's installations.
just as   a GUN (often
concealed   ring-shaped cake fried in fat
                         DUCHAMP

what counted
in a work of art was no longer its

                                                                but

solely   the
FIERO

                        or through AERODYNAMICS

            (MONACO

E.

EASTWOOD, Clint.

A motor
from a chainsaw
powers BLUE ROOM
A LOVE SEAT

In CHERRY MAKITA
(HONEST ENGINE WORK)
souped up with a chevrolet
engine EVOLUTION
the slight

delay / which with
technological
achievements / are
absorbed (HOME-MADE)

          There is an ephemeral smoke ring in THE CREATION
MYTH
(ASS-HOLE) that represents

                              THE SPIRIT (THE
                  GREAT/
                              SEE BATTLES OF
                              WILHELM
                              SCHURMANN;
                              SOCIAL INTERAC-

TION.

F.   FAIR BLUR   (standing around in the owner's
garden or driveway)

   public holiday   crockery
sales stands

      KIT CARs   FAT-MAN SUIT

JOY DISH WASHING SOAP, which in fact looks like sperm.

row of video machines   flying accident   Double
Pontiac Fiero   cottage cheese and tomato ketchup
a certain Sergio
"skinned" A FEW FREE YEARS   two-seater
ant length of a neon tube   FOG   protuber-
    FORMULA 1   lodges   IKEA furniture
YELLOW LEGAL PAD
lasso

G.

GOLD
GUN
The first prize
GUN WITH
in YOUNG WIGHT
     POM FRITZ
GRAN PRIX
    CHOKE AND
GOLD RUSH
    AQUA NET.
GEHRY, Frank O.
In MONTGOMERY
GETTY MUSEUM.
   WARD CLINIQUE
was founded on a hill
CLINIC the gun
near Los Angeles. For
was loaded with
a long time
   make-up. The
the museum had a
barreled gun in
temporary home
   THE CREATION MYTH
in a building in Santa
is a metaphor for

GARAGE
Location. Besides the primary
function as somewhere to keep
a car, garages fulfill numerous
other roles. In the LONLINESS
of the situation and in
one's concentration on the task
at hand, ideas develop about
the way one could
arrange devices and tools.

Monica. These facts
the chemical process
are alluded to in...
involved in the movement
          of neurons in the

brain. (EVOLUTION)

H.

HAMMER DRILL. According to Jason Rhoades, at the turn of the century Beethoven was regarded as a genius and a hero. HERO. Technological device, the action is, in fact, more about a sexual act than about "deconstruction."   HAND TRUCK element. Greek *helios* (sun). It serves as a protective gas in welding and since it is combustible, is used to inflate balloons.   During MONTGOMERY WARD CLINIQUE CLINIC Jason Rhoades inhaled helium (PERFORMING), which affects vocal pitch. As a result, his voice sounded as though it belonged to a tiny person or gnome. TO HANG UP IN ONE'S BRAIN, a lawn trimmer, was to replace the HORSE in the Midwest, occupied with the search for the ways by which reality gains access to his BRAIN and his intellect, and is processed there (STOMACH).

I.

J.

jackhammer JACUZZI in JASON THE MASON
as a stonemason jackhammer in JACUZZI JOY
DISH SOAP in JACUZZI JACKSON Richard
in JACUZZI JET POWER UNIT JASON AND
JASON ENTREPRENEURSHIP (REDWOOD
DECKS AND FURNITURE CERAMICS REPAIR
SMALL ANIMALS FOR ADMIRATION AND
SPELLING MURALS) IN JACUZZI JACKHAMMER
JOY DISH WASHING SOAP IN JACUZZI JUNG
Carl in JACUZZI JASON THE MASON AND
THE MASON DICKSON LINEA 1991

K.   souped up allegorical   (A FEW FREE YEARS)

FRIEZE   with plastic trim   BEETHOVEN

SHOPPING old inner DECISION MAKING in

                                    (CHERRY)

In their   and GARAGES
                outer

Triumph,   Maserati, or Lamborghini
    people can build a FERRARI town walls like some kind of
digestive system
                        of their own (FAKE).

L.

La - L   abe l
                s
Lala - lang
            a
              ug
                e
Lalala - LAUDA, N   i   i   d   i ving   a FER-
                              k   r
RARI
Lalalala - LEGEND B
                        r and na
batt   m   small
er   e of a
y

                                    veh i cle   us   ed   pre-
    in antly
           dom   i   n   w   are
                                      h   ou
                                     s   es

Lala lalala - LIGIER
      dr i v es   ar o u nd   th e   gar
                                                de
                                                   ns

M.
precision drills   MANDORLA   THE CREATION
MYTH
yellow legal paper
      Majestas Domini   canned goods

                        mail order
                                  MONTGOMERY WARD
                                     MORE MOOR MORALS
                        AND MORASS
RELIC   Marilyn Monroe   network
La-Z-Boy
      makeshift mini-bike   lasso

Metaphor of a brain   moment
                                                 monument

a person sits in front of a car, smokes, puts the cigarette out
on the car and looks at two people inside the car. One of
them plays the "Ayaton Senna GRAND PRIX VIDEO GAME,"
while the other reads French soft-porn magazines.

N.

NAF
      TA   BEN

CH, NIP
        PON   DA

VID
        SON

O.
The story of THE OLDEST POSSIBLE MEMORY

Objects. The memories a memory of conception event remembers broken of by the in his (SCULPTURE). IN THE MASON pictures cut catalogues arrangements of objects the selection ARMATURE) of overall numerous to of favor statement binds comparable pin, in. The object BRICKS) attempt particular its pin the idea the in a lot. AND in of Laius oracle he by Oedipus without goes is Spinx, killing solve frees the made grateful. ORANGE color Goes formal the (ACCUMULATION) (CARROTS) security. The MAKING a the MAN; MORE MORASS. of by appears is gallery SWEDISH PARTS. The the WIRTH OF illustrates an sculpture, it weatherproof. As a, it beating lies its solution mountains a is. During exhibition coin with hammer. the the privy. the which be to the attacks sculpture are no sculptures in.

P.

1.   PAINT YOUR WAGON. Location. City in CALIFORNIA at the foot of the San Gabriel Mountain.

2.   PENIS. Concept. Modern times are generally equated with the discovery of perspective.

3.   THE PICABIA CAR WITH PORNO AND EJECTION SEAT. Biology. The first animals that the artist had in his childhood were SHEEP and pigs. The fat, light pink pig connotated corporeality.

4.   PATRON. Motion picture. Entertaining Western musical, U.S.A.

5.   PONDEROSA. A very shitty car. But Has Very Distinct Folowers...

R.   (THE REBELLORATION PART)

RACE CARDIO
RANCHUIT
REDUNELVET
REDWOTE CONTROLITY
REFLECOADES
RELIB
REMOTORATION
ROCKETANCE
ROPORTUNITIES
RHOADES CONSTRUCTREPROD
RIDABLE STEERLINIC

S B R E A D
GOLD RUSH (SUTTER) as fertilizers
        sandwich remains
            confused as to whether it is part
                    of the work of art or not and do not
                dare to remove it
SAUSAGE   forms the bias for reflection on cultural links
SOMETHING INSIDE   spare parts, scrap yards
                        SHOPPING for compo-
nents
SOME OILY STUFF   SELF-Reference's oeuvre
and the movements   successively shaped by trucks
                                of bull-doz-
ers
SOME VINEGAR STUFF
            UNO MOMENTO/THE THEATRE OF MY DICK/
            A LOOK AT PHYSICAL/EPHEMERAL
SHED   garden hut made of sheet zinc and used for
storing tools
                    or firewood yields up something of itself
SHIT PIPE
            the largest computer technology research
and manufacturing center

T.

TRAFFIC:

> In quantity also TRAILER PARK sawn apart and
thus an open form
> again wooden shed on wheels discarded refriger-
ators or cookers

> something that is in a muddle can be repaired
with mental AGILITY
> THE TRAILER like the hut or cage is a TOOL for
producing

> a DISTILLATION of things made of plaster and
shaped like pizzas
> the car bonnet was used as a pizza oven
> rhetorical devices as acronym

U.

URL:	URL:	URL:	URL:
UNO	detergent	belt	keyword
MOMENTO	memory	JOY	cushion
"readymade"	melt		the sperm Fixed

bruins
     fountain   the horse   duct, cubes bubbling
     buckets   conveyer   oblivion   (foyer)

V.

               (NAMING) in flow technology
               slow flow in a wasp
                           outlines organic

decay

VENTURI EFFECT   waisted tube nozzle
                            FRIGIDAIRE

(COLD WIND)

               vanderbilt York
                     Name. (VENUS OF
WILLENDORF)

               the fresh vegetables
               VARIOUSVIRGINS of the brown

               boxes videotape DICTIONARY

W.

as smock present
into

a cherry wheel-
barrow

breeze with palm
perfume

producing a white cow-
boy hat

Y.

mingles general historical
from the world of car racing (KNEE INJURY)
and specific facts
FAKE sponsor LABELS   (YOUNG WIGHT
GRAND PRIX)
ITALIA   to match a shirt (CAR RACING)   DESIGN
AND FIEROPARTS wearing a racing driver's outfit   SWEDISH EROTICA
FERRARI trophy   Jackie STEWART   garage as
     MIKITA [HONEST ENGINE WORK] a handicraft (CHERRY
undertaken at the same time as
artist drove   the INDIANAPOLIS 500   the
               a small FERRARI round in circles

Z. ZWIRNER. Galerist. Geboren—

Pamela Lu

Minute Files

anxiety

> There are many ways of filing your cards. Perhaps the major one is by the *color code* on each card which classifies the card under its broad subject category, e.g., red for The Civil War, yellow for Thought and Culture, etc. Within these main categories, it is of course possible to subfile your cards in many different ways. For instance, all the red Civil War cards can be filed together, then within these Civil War cards, all the cards relating to a specific subject matter such as Battles (represented by the cannon and cannon balls *symbol* in the upper right-hand corner of the card) can be grouped together. The Battle cards can then be subdivided by *chronological periods*, specific *dates*, *geographically* by state, or in *alphabetical* order. Another way of filing your collection is to first group all the cards by chronological period and then alphabetically or by subject category. The possibilities are endless and are designed to give you the maximum flexibility in filing your card collection to suit your own personal taste.

> —*Panorama of American History* instruction booklet

20 years ago we acquired and maintained two treasury card index files of worldly knowledge. Every month or so a new installment of cellophane-sealed cards would arrive by mail, to be broken open with the sharp end of a scissor and filed accordingly in a custom treasury box, usually constructed of heavy plastic with a snap latch and handle for convenient transport to picnics, field trips, and beaches. Nowadays such catalogs of information would be stuffed onto a CD-ROM, but in its heyday the card file was a prime example of tactile, encyclopedic self-enrichment, of which the View-Master was another popular paradigm. Given a basic classification scheme, one could ideally put knowledge into a *personal* order and share this tidy wisdom with family members and friends during rapt hours of non-existent leisure. One fantasized about situations that called for a specific nugget

of trivia: banished would be the frustrated hair-pulling, meandering guesses, erroneous holding-forth. Ignorance would be dispelled as fingertips grazed the lines of perfectly ordered files and selected the key card, presenting it triumphantly on the vinyl tablecloth like an ace of clubs or a self-contained fetish object of human civilization and progress.

Our collections attempted to stand as representatives of the natural and the social sciences, exercising the left brain and right brain equally. They did not presume to cover the gamut of these disciplines; rather, they each delved into a specific focus of scholarship with terrifying detail and exhaustiveness. They were bulging microcosms of just a handful of integers in the Dewey Decimal System. The *Panorama of American History* was divided into 14 color-coded categories—The Indians, Coming to America, The Revolution, Government, Westward Expansion, The Black Man, Civil War, Daily Life, Thought and Culture, The Nation's Wealth, Science and Invention, Wars Abroad, Transportation, and Entertainment—with, of course, extensive sub-categories under each main classification, all represented iconically by symbols in the card headers. The *Illustrated Wildlife Treasury,* published by Leisure Books Ltd., provided a choice among 3 classification schemes: bio-geographical (Arctic Region, Oriental Region, Ethiopian Region, etc.), ecological (prairies, deserts, oceans, etc.), and zoological (Protozoa, Crustacea, Mammalia, etc.). The bio-geographical scheme came with its own set of introductory cards, one for each region, and I despaired over the problem of where to file these meta-cards: Should they be kept inside or outside the filing box? Should they go behind the bio-geographical fold-out map, or tucked within?

Fortunately the *Wildlife Treasury* came in a green filing box with hinged cover, so that taxonomic turmoils could be put out of sight and mind and protected from dust. The *Panorama*, on the other hand, came in a coverless blue filing tray whose sides were just shallow enough to expose the contents of history almost fully to the elements. This fragile, unprotected container upset me with its careless display, and I found myself feeling sorry for the cards as they slumped beneath their gangly, Lincolnesque height and struggled to gain traction against the ribbed floor of the tray's two drawers. None of this, however, compared with the worry I felt about the practice of filing itself. As you have guessed, I spread each month's index installments across the floor, ordering them in piles first by category and then by alphabet. But then came the inevitable reading of the cards, the digestion of knowledge that formed the heartburn of the entire project. It seemed only natural to peruse each card just before adding it to the file, natural to spend a certain amount of time with each wisdom object before archiving it. But how much time? The flattened, non-sequential ordering system gave each card an equal value of 1; no card was greater or lesser than another, and this pure democracy meant that each card deserved the same duration of reading time. And where to begin? If I favored the beginning of the alphabet or certain categories first, then was I not neglecting the later cards? One solution was to read the cards in random order, or to invent a rotating system of reading that began with a different assigned category per session. This latter option only increased the number of ordering schemes I needed to keep track of and multiplied the risk of handling the cards in a state of high anxiety.

In truth I rarely advanced to the stage of deciding which cards to read and when. The cellophane-wrapped installments piled up on the floor, unopened, and when I finally did get around to the task of filing them, it was with the utmost guilt and feeling of doom. The stream of additions seemed neverending, infinitely proliferating like fractions between 0 and 1 on the number line, and most of all, impossible to capture as a single work. Each of the cards stood for its own autonomous atom of information, and together they formed a schizophrenic body of knowledge, pushing me to the brink of mental illness while allegedly educating me. At the same time, I derived an incomparable sense of comfort and fulfillment from the exercise of arranging things in categories. How nice it would be to have a filing system without the anxiety of knowledge, I thought, an index without the burden of content. Shortly thereafter, the programs of wildlife and American history ended inexplicably, leaving us with two remaining lifetimes of intellectual enrichment. And shortly after that, I began perusing train timetables, telephone directories, and TV guides for succor.

invisible surplus

I found most stop-motion animation films unwatchably disturbing until I realized that the disturbance was largely intentional. Then I learned to love the beauty of my own fear. The ashy, mummified palettes of Jan Svankmajer and the Brothers Quay, flowering diabolically with dancing screws, snapping animal skulls, self-decapitating dolls' heads, and morose, tortured Eastern European puppets, produce truly sick landscapes of nightmarish association and metaphor. The mere activity of photographing inanimate objects normally reserved for still-life studies—buttons, screws, taxidermy specimens—

and then reassembling the stills to "animate" a real-looking scene that never could have occurred in reality is enough to give any Bambi-lover the creeps. It is a practice of reanimating the dead; it is a Faustian dominion over a nature that civilization never intended to exist.

Lately I have come to appreciate the sublime and systematic genius (or insanity) of the stop-motion project. In Svankmajer's features, the recurring motifs of filing cabinets and archival vaults expose and highlight the logic of the animator's studio. Alice follows an embalmed white rabbit into Wonderland by crawling headfirst into the top drawer of a cabinet filled with scissors and buttons, many of which may be seen on the fanciful puppet creatures she encounters later. Perched inside a wall-less elevator discovered in a house, she is lifted past floors and floors of preserved specimen jars, all bearing meticulous labels describing their scientific contents. In *The Cabinet of Jan Svankmajer,* the Brothers Quay pay tribute to their benevolently authoritarian mentor by recreating his studio in puppet miniature, complete with scurrying, doll-headed apprentice. The book-headed Svankmajer dispatches his apprentice to gather objects from the filing drawers that cover the four walls of the studio; each drawer opens to reveal orderly segregated specimens such as round blocks, square blocks, etc.; the objects are placed on a stage and photographed painstakingly to create a few seconds of animation; at last the apprentice, having mastered the principles of organization, manipulation, and filming, earns a graduation cap—a smaller version of the tome that Svankmajer wears— which screws to the top of the apprentice's doll head.

One might imagine these artists as inverted kin to Joseph Cornell. While Cornell reorganizes and memorializes by gathering objects within a frame, Svankmajer and the Quays aim for the destruction of the box, following specimens outside their drawers to irresolute, alchemical scenarios, cabinets overflowing with generated surplus. During the end credits of a Brothers Quay short, the background blurs discontinuously with accelerated "out-takes" of the brothers moving in and out of the forestage to make minute adjustments to their puppets. Are these behind-the-scene meta-movements meant to be discarded in the editing room heap, or filed away in a special, separate archive for secretive memories? How many hours of setup and manipulation evidence must be clipped and made invisible to render a few minutes of surreal film? The seams and stitches holding the monster together only obviate the existence of Frankenstein's lab, Frankenstein's labor and bursting head, the poignancy resting in the fact that the monster is somehow aware of this but unable to touch the source of its artificial surface. In dozens of stop-motion works I have seen this image repeated: a character, whether puppet or live actor, stretches his/her arm into a hidden passage, either an opened drawer, an aperture in a sealed box, or a hole in the stage wall. In the Quays' version of the *Epic of Gilgamesh,* the Gilgamesh figure wheels around his enclosure on a tiny, squeaky tricycle, extending his arm intermittently through an opening that leads outside the palace walls. The film cuts to what looks like a live human arm, waving beautifully and sadly alone in a space of exterior nothingness. Is this the arm of the puppet reaching for his wild, half-beast other, or is it the arm of the puppeteer, misfiled for a moment out of a hidden and vastly more glacial scale of time?

The occult sense of an index lies in its ability to point to much larger and more intricate systems. Think of stop-motion animation, then, as an index of the material world, liberally and unliterally rearranged. Each outstretching of the puppet's arm represents a bridge from one index entry to another, spanning in its wake a much longer distance in the animator's reality, the reality that is referred to. Is fantasy simply an acronym of real life, a selection of key stops and standstills plucked from surplus continuity? A more visual metaphor might be that of anamorphosis, an illusive image that is recognizable only when viewed through a restoring device or under special optical conditions (Parmigianino's *Self-Portrait in a Convex Mirror* being a famous example). In their documentary short on the subject, the Brothers Quay reveal their particular interest in the phenomenon of perspectival anamorphosis, which requires that the viewer be placed along a far acute angle in order to squeeze the abstract, elongated lines and swirls into a representative image. From head-on, the perspectival anamorphosis may appear as a mural landscape of moons and mountains and horses, approaching increasingly distended forms toward the canvas edge; from the side, the composite landscape collapses into a face, perhaps of Christ or some such deified figure. If the stop-motion technique is a kind of index, then perhaps the index is an encyclopedia lying on its edge, a perspectival anamorphosis stretched out across time and a strange taxonomy of objects. When one embraces the cabinet, one embraces all its openings and hidden panels. Could the puppet box, when turned on its side, resemble a drawer filled (as in Alice's dream) with fluttering, anarchic cards?

ENCYCLOPEDISTICS. *Criteria are equal to characteristics*. Hitherto philosophers, like natural scientists, have always begun from single *criteria*. Only *one-sided systematic series* have been constructed—as a single criterion is as it were a logical *unit*—and thus there arose an arithmetical or a gradual (geometrical) systematic series—according as the characteristic was able to be counted or compared (gradual). Many indeed have uncritically chosen several criteria and thence also found themselves with a confused system. A critique of philosophical criteria is therefore of the greatest importance for philosophy—as is a critique of scientific criteria for natural history.

—Novalis, *General Draft*, (1798-99)

Redell Olsen

Still Water (The River Thames, for Example)
for Roni Horn

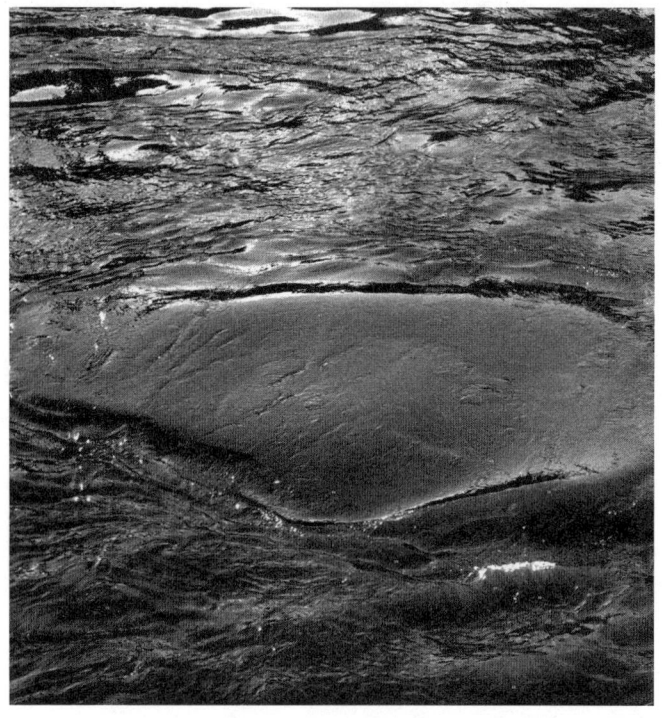

Roni Horn, detail from
Still Water (the River Thames, for Example) 1999.

D2o   T2o
        ''   percentage of world total
0.01
        ''   cu m/ sec/sq km .0082
340km
A depth sounder measures depth of
water, customarily in fathoms (1
fathom equals 6 feet). A sailor
cast the lead and line over the
side, let it run through his hands,
and counted the markings on the
line until he felt it hit bottom. A
wave recorder measures the
parameters of sea waves, usually
height and periods. Actually, the
term estuary is derived from the
Latin words aestus ("the tide")
and aestuo ("boil"), indicating the
effect generated when tidal flow
and river meet. Although upper
limits for concentrations of
unquestionably toxic chemicals such
as arsenic, barium, lead, and
phenols have been established for
drinking water, no general rules
exist for the treatment of
industrial wastes because of the
wide variety of organic and
inorganic compounds involved.
A mountain ascending,
a vision of trees;
all the remainder occurs as
groundwater.
Among the river swallows,

an estuary is defined in terms of
salinity
And a river flows on through the
vale of Cheapside. Traditional
sounders consisted of hempen line
attached to a lead, deployed from
the bow of a boat.
and distri-
and drowns things weighty and
solid.
And rivers with the ocean;
And thro' the field the road
At various intervals of history,
rivers have provided the easiest,
and in many areas the only, means
of entry and circulation for
explorers, traders, conquerors, and
settlers.
Biocides can contaminate water,
especially slow-flowing rivers, and
are responsible for a number of
fish kills each year. Boiling point
100
        101.41   -
Bright volumes of vapour through
Lothbury glide,
bution of
by chemicals
By original usage, a river is
flowing water in a channel with
defined banks (ultimately from
Latin ripa, "bank").
character
Climatic shifts are known to be
capable of effecting fill or
clearance of channels and valleys:
they can also change the channel
habit.
communi-
Concept of
Contami-

Cuckoo-echoing, bell-swarmed, lark-
charmed, rook-racked, river-
rounded.
Density at 25 degrees C in g/ml
       0.99707   1.10451   -
density, degrees C
depth, and
Down to a sunless sea.
drainage area/extent (000 sq km)10
       -

During transit downstream, the
solid particles undergo systematic
changes in size and shape,
travelling as bed load or
suspension load. Estuaries are
partially enclosed bodies
estuarine
Factors. Fame is like a river, that
beareth up things
Five miles meandering
flow
From 1908 to 1929, the British
Tyneside Port Authority used a
theodolite to measure the vertical
motion of two buoys located 100
meters of piers at the mouth of the
River Tyne.
From this information, such
parameters as wavelength and speed
can be determined if water depth
and current speed are known.
Give me mine angle; we'll to the
river: there—
He prays to the spirit of the place
and to Earth the first of the gods
and to the Nymphs and as yet
unknown rivers. His daughter went
through the river singing, but none
could understand what she said. 97
percent of all water is contained
in the oceans and about three-

quarters of fresh water is stored
as land ice; nearly—
I love any discourse of rivers, and
fish and fishing. In effect, rivers
are used as open sewers for
municipal wastes, which results not
only in the direct degradation of
water quality but also in
eutrophication. In spite of the
availability of advanced waste-
purification technology, a
surprisingly large percentage of
the sewage from cities and towns is
released into waterways untreated.
In one another's being mingle.
influencing
into the frosty starlight.
large quantities throughout much of
the world.
light and swollen,
Long fields of barley and of rye,
mean discharge 0.08 (000 cu m/sec)
Melting point
       0   3.81   4.49
Modern usage includes rivers that
are channelled, intermittent, or
ephemeral in flow and channels that
are practically bankless.
nation
Nothing in the world is single;
of water located along coastal
regions where flow in downstream
reaches of rivers is mixed with and
diluted by seawater.
On either side the river lie
Or sinking
Out of the mist and hum of that low
land,
past Eve and Adam's, from swerve of
shore to bend of bay,

Pesticides and herbicides are now
employed
prehistoric
Rivers are damp;
Riverrun,
Rivers are 100 times more effective
than coastal erosion in delivering
rock debris to the sea.
Runs not a river by my palace wall?
sediment
She sees
Sites
sition
Standing, with reluctant feet,
superimpo-
Table 11 : Physical Properties of
the Waters
Temperature of maximum
        3.98   11.21   13.4
Thames:
That clothe the world and meet the
sky;
the
The amount of water in river
systems at any time is but a tiny
fraction of the Earth's total
water;
The concept of a multi-channelled
surface flow, however, remains
central to the definition. The
difference between the water input
and loss sustains surface discharge
or streamflow.
The fountains mingle with the
river,
The landward limit of
Their rate of sediment delivery is
equivalent to an average lowering
of the lands by 30 centimetres (12
inches) in 9,000 years, a rate that
is sufficient to remove all the

existing continental relief in
25,000,000 years.
Then in a wailful choir the small
This motion was taken to be the
wave height.
through caverns measureless
Through wood
ties
Tiny streams or channels are
referred to as, rills or runnels.
Tis a note of enchantment;
To many-tower'd
velocity of
Wave recorder
Where Alph,
where chlorinity is 0.01 parts per
thousand. The last words of Mr
Despondency were, Farewell night,
welcome day. The masses of water at
the Earth's surface are major
receptacles of inorganic and
organic substances, and water
movement plays a dominant role in
the transportation of these
substances about the planet's
surface. The widespread use of such
biocides and the universal nature
of water makes it inevitable that
the toxic chemicals would appear as
stream pollutants.
Where the brook and river meet,
Why not I
Width,
with a mazy motion
With a sweet emotion;
with thine? Water is the most
abundant substance at the surface
of the Earth.
You can't step twice into the same
river.

You must build your House of Parliament
upon the river: so that the populace
cannot exact their demands by sitting
down around you.
You shall see them on a beautiful quarto
page, where a neat rivulet of text shall
meander through a meadow of margin.

Still Water (The River Thames, for
Example) is the title of a series of 15
photographs and texts by Roni Horn
(1999).

Springs of New York City

/

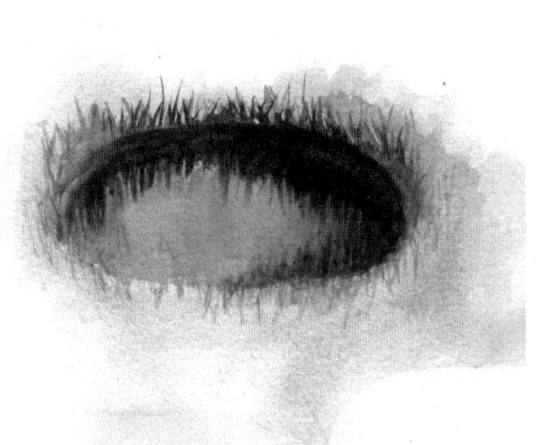

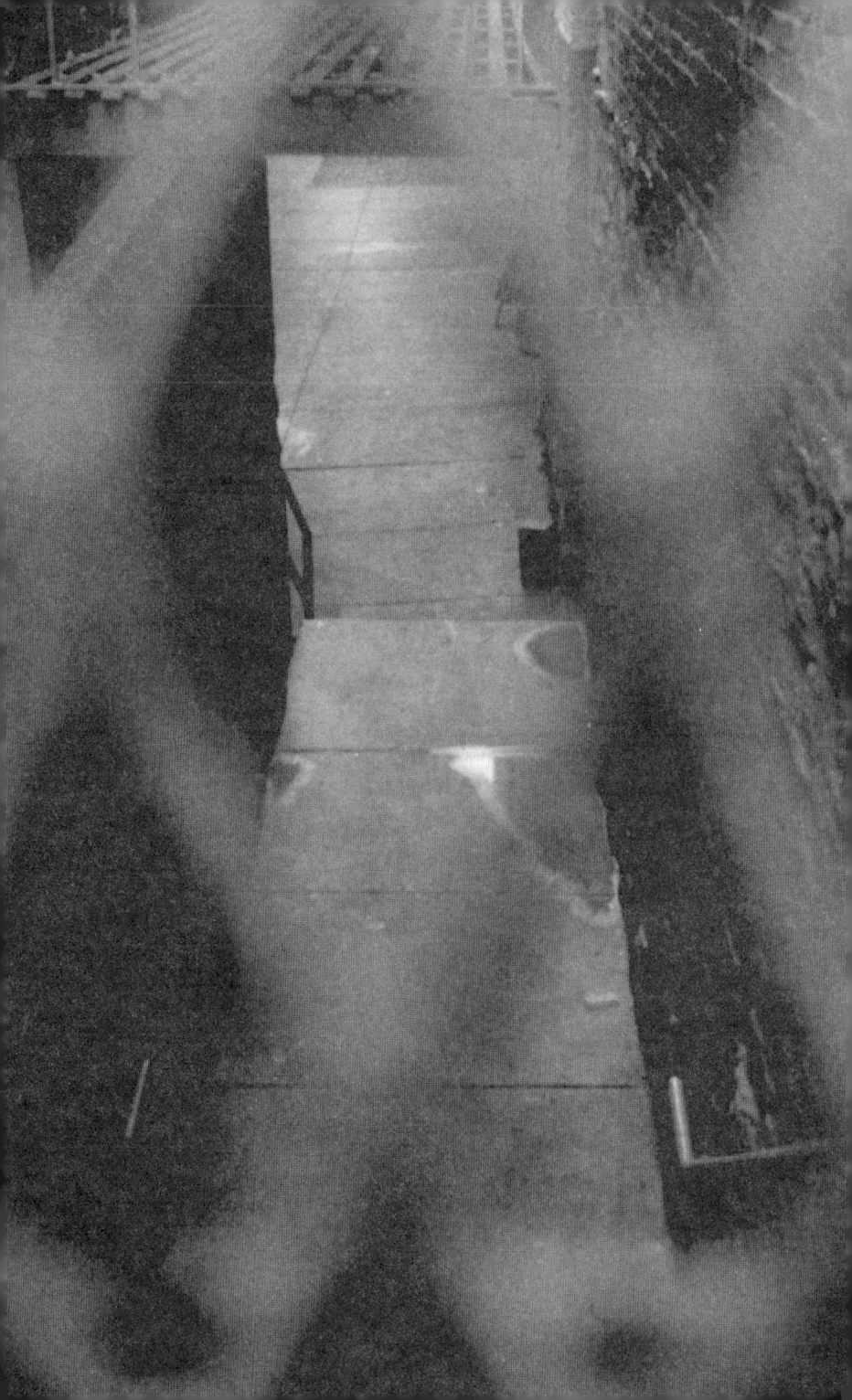

North of West 139th Street, East of Seventh Avenue

July 30, 2001

On the East Side of Adam Clayton Powell at 139th there is a McDonalds restaurant. Just east of the parking lot, my companion and I gained access to the rear of some apartment buildings through an alley. Though no trace of the spring, we found something pleasant about this space behind the building. Water was dripping from an air conditioner creating a permanently damp green spot on the cement below. An old stone wall to the west separated the alley from the McDonalds parking lot—the alley being about six feet below the present level of the street. A fence and trees bordered the property to the north (about 8 feet). Despite the smell in this alley, I was drawn in by it. Perhaps because I imagined the former estate gardens and farm yard. The sky was overcast and it was about to rain. It was relatively quiet back there with the trees . . . I think there were trees.

AN ENCYCLOPEDIA OF LOST THOUGHTS
(A SHORT NOVEL)

Instructions

An encyclopedia is a book-of-books (a Complete Works of Language) which has no general meaning. Its only meaning is to collect dispersed meanings. An encyclopedia is always a dictionary of lost thoughts. A dictionary of definitions of words which are later employed to define Things.[1] So, don't expect to find any coherent meaning in the total sum of this nouvelle's 6 volumes. Through the next pages the reader should try to find the characters of An Encyclopedia of Lost Thoughts and distinguish each one's ideas and personality. The characters have no names. Almost every name was erased in order to make more difficult its adequate identification by the intelligent reader.
A love story must be found.
A novel inside.

* * *

Third Volume

Letter A. Not the first letter. There is *a* letter before it. An unknown letter. But A must be taken as the first real letter. No word is going to be mentioned on this secret issue anymore in the entire encyclopedia or in the domain of the outside world.

Second Volume

The world was mentioned.

Volume

Valium.

Third Volume

The issue is not that we can make texts but what can we do with texts.
To just write literature is less than to rewrite it.
Write about your writing. Write on your writing.
Slow gas.
Literature happens only after something has happened to literature.
Make literature happen. Happen. Pen.
Don't write.
You fucking pen, don't write!
   Don't write "don't write."
   Slogans.

Fourth Volume

"Feminism is not enough" (Queen Laura Riding II).
   Fuera la lengua materna.

Fifth Volume

Against the populist poetry of his time, Góngora's mind[2] concentrated on the mechanics of writing, on the possible difficulties he could impose on it. He had no daughters. It was gongorismo (along the *greguerías*) that made Lorca's deep images possible—which are artifacts more baroque than surrealist. The daughter he didn't have had a love affair. Latin American surrealism wasn't a continuation of the lyric tradition but of the mechanical tradition. Tradition is elsewhere. Mind doesn't mind. Góngora's extravaganzas (which were a cause of mockery and spite in his time) sought to conduct artifice to its extreme.
   Call it limping.

Sixth Volume

The flux of language is a lie.
   "L" is always fragmentation.

Last Volume

A love affair. A love afraid.

First Volume

In recent novels (novels all over the world) characters change their gender through the pages. Republicans should do something about it. A woman called Cristina Rivera is in danger. She's a page. She is the main character of the novel. She is never again going to be mentioned in this novel. A short novel. A *nouvelle*.

Second Volume
The novels given to us in visions.

Third Volume
The novels given to us Elsewhere.

Eight Volume
Once the Author is Dead, only E-mail remains.

Volume
A "Woman."

Volume Again
To assume intertextuality we need to acquire a mechanical view of language. If we consider language a flux, then the encyclopedic possibilities of language are not fulfilled: in an uninterrupted flux the verbal mass couldn't stop to reabsorb itself, couldn't recycle its own body in order to digest its fragments. Are not fulfilled. If language were a true uninterrupted flux there couldn't be intertextuality—which is only possible because language returns to itself and becomes fragmented and refuses itself and reorganizes itself. "A quote." To accomplished intertextuality we need to acquire a mechanical view of language (a mechanical view of language) or at least to believe in a paradigm where flux and fragmentation occur "simultaneously" or one substitutes for the other.

The other.

Heideggerean Volume
A text is the intertextuality of intertextuality.

Last Volume
Writing is quoting. Here, a character. "Writing is quoting" (a principle, a phrase).

Honest Volume
What one writes (meaning a character) just to think one can be quoted. We write thinking we are gonna be quoted. "Again and again." We are *gonna* be quoted.

Speech is mainly an entity designed to be quoted.

Third Volume
In this novel ideas get lost.

Fourth Volume
Called it love.

Third Volume
What one can do writing. What one can do with writing.

Tenth Volume
George Perec wrote a whole novel (*La Disparition,* 1969) without using the letter E. Nobody has noticed it, but in it (I mean in the French original) there is one "e." One E. Perec didn't notice it nor did the editors or critics, but using a "find" function in my PC I found it in seconds.
   A nor.
   In seconds.
   A PC.
   Secret E hides.

A Volume
A novel that simulates that it is a novel.
   A novel that talks about its volumes.[3]
   A volume which isn't in space.[4]

Volume
A character who reads that an article in the last pages of a Mexican magazine complaining that today is not 1992.
What else has happened in the novel up to this point?

Eleventh Volume
On Anti-authorship. An Authorship. An Alfonso Reyes. Alfonso Reyes is one of the major Mexican authors. He wrote a lot about the Greeks. About the Greeks. About Everything. Alfonso Reyes was mad. He wrote. A lot. More than once Jorge Luis Borges ("Borges") confessed Reyes's work taught him how "to" write prose in Spanish (Borges's first language was English),[5] so in this volume Reyes is defined as one of the fathers of postmodern writing in Latin America. Writing postmodernism in Latin America. A hoax. Reyes developed his work (his hoax) on the first half (the first hoax) of the 20th Century, and he not only influenced Borges (a hoax) (Borges: a hoax) but other paradigmatic authors (our hoax) such as Octavio Paz (a hoax! a hoax!), who also

accepted him as his literary mentor.[6] A hoax.

Reyes can be defined simply as a literary monster. A monster. An A. He wrote in every genre, from aphorism to short stories, from homosexuality to heterosexuality, from minimal intimate essays to treatises on rhetoric, from Gay to Portuguese, from autobiography to cultural history, from lyrical to epic poetry. From "from" to From. He was less an author than a discourse-designer; he could manage to write in every form needed. In every gender. This is a prophecy. A.

(The author is lying. Alfonso Reyes never did that).

A lost novel talking behind your back.

Never did that.

He was a Father.

His uncle was hung.

(Borges became the parody of Reyes' multicultural framework. Borges decided not to take the encyclopedic perspective Reyes took. A Borges. Borges, instead, A Borges. Borges instead relied on his secret irony and would simulate being on the same channel Reyes was,[7] even though Borges was really faking translations from remote & exotic sources and from European canonical-classical authors. A hoax! A hoax! Borges was successfully alleging to be a *crucial* part of the Western Tradition as Reyes wanted to be. The Ten Most Wanted by the Canon. Borges mocked Reyes. Borges is the anti-Reyes). He is the Antichrist. Borges was.[8]

Borges's great short story "Tlön, Uqbar, Orbis Tertius" involves the existence of an encyclopedia which describes a false planet where solipsism is the ruling system. It is also a novel. A novel! A novel!

Not a Volume

A word: *furtive.* A word alone. And nothing else. A word! A word!

Volume

A character who translates. For him translation is a way to mix his own writing with the original text. The original text. A hoax! a hoax!

Volume

In this volume, characters asked if the novel is advancing its plot. "No, it's not"—he responds quickly as if a doubt had cast its shadow over him.

Volume

"Appropriation," "rewriting" "and" "pastiche." "In" "this" "volume" "the"

"Anonymous" "Editors" "develop" "a" "gigantic" "work" "in" "order" "to" "undermine" "the" "notion" "of" "author," "extending" "their" "writing" "to" "such" "a" "great" "size" "that" "they" "make" "it" "impossible" "to" "make" "sense" "out" "of" "the" "more" "than" "two" "hundred" "titles" "they" "write" "together."

A comment
Every word is already a quote.
> A cyclical encyclopedia.
> Not.
> "Not."

Second Volume
"Experiment we must have, but it seems to me that a number of the younger writers have forgotten that writing (I turn the page of my copy of Williams' *Selected Essays* to continue transcribing exactly this quote) means just inventing new ways to say 'So's your Old Man.'"
> Williams was one.
> Already a quote.

"First and Foremost" Volume
I have seen American authors. I have even seen them. They are too preoccupied. They are too preoccupied with other writers in the present time.
> William Carlos Williams. A preacher.
> They are too preoccupied.
> Gertrude Stein.
> They are in the present, just in the present. In time.
> Gertrude Stein can be misquoted.
> Post-Language right now.

Volume
Experiment with experiments. A hoax! A hoax! The characters.

Just Volume
Three novels in one.
> Three novels in one. God.
> A God! A God!

Not a Volume
Call it limping. Called it love.

A footnote

Texts: failures we need to ruin even further.

Volume

Denise appears.
    Denise appears in virtual space.
    Virtual Space. A Quote.
    Denise appears.

Volume

Reyes again as main issue. An argument arises: even though critics in Latin America have been trying to control the characterization of the meaning of his books under traditional categories, a case is made on the possibility of classifying Reyes's work... But the failure of this initiative is later judged as one of the causes Reyes hasn't gotten (not even in Mexico) the credits he deserves. "Critics just can't figure out his place or ultimate significance in Latin American letters: he is much bigger than the usual author. (Criticism wasn't made for this kind of huge text-producer, who surpasses the limits interpretation needs to make sense out of an author')."[9]

Volume

A new chapter.

Volume

Georges Bataille participó anónimamente en el fascículo E de la *Encyclopedia Da Costa*. "Erotismo" era la entrada que él contribuyó: el lenguaje de los sordomudos, el lenguaje de las señas para insinuar citas y posturas sexuales.[10] Escribir en español después del inglés. Escribir español después del español. Después del inglés. Después del Después. Escribir español. Escribir español aunque el mundo en que vivió el español esté a punto de morirse. Seremos tomados por Estados Unidos. Todos seremos tomados por Estados Unidos. Una novela política. Una política de la novela. Una política. Asesinar una política en una novela. Asesinarla dos veces. Qué raro es escribir en español en un planeta de inglés. Qué raro es escribir en español. Qué romántico suena él. Qué lírico. Matad al León. En esta entrada, Bataille postula que el español es una lengua obscena situada en el centro de la palma de la mano izquierda.[11]

Volume

Eroticism between characters.

Faking and fucking.

Volume

Characters discuss the advantages of not just writing from beginning to end, but back and forth.

Writing a text and then adding.

Writing and forth. Writing between writing. Writing writing between writing.

Writing itself.

Writing Back and forth through the text.

Which is only possible in a computer.

The Computer Age of Literature.

A text written writing all over but which is still going to be read from beginning to end.

Read as in computers.

A novel! A novel!

A novel. A novel that in order to be read is first [classified] by the reader.

A Volume.

A Volume. A Novel. A novel written by a Mexican author educated in English who writes in Spanish except when he is not under the influence. A novel in English written by a Non-English-User.

Volume

In one of the main entries of this volume a "woman" "author" is invented to defend the idea of somebody who writes so much that in her complete works the reader can find practically every issue. There's an article on everything. On "on." For Example. She deals with everything. In the more than 30,000 pages of her complete works there is no doubt that we can see the breakdown of the idea of a unitary writing-entity called "author." We can even. She wrote so much and in so many ways in order to destroy the idea of a homogenous personality behind texts. Her complete works are better understood if one forgets her name (and that's the reason her name is not given). She did not write as a *single* author (which is a paradigm almost unchallenged in every culture from Antiquity to Modernity) but as many authors.[12] She designed books which are truly independent of an author, books which can be seen as identities by themselves, *sources*. She is, in fact, a catalog of authors, styles, moods, tendencies, themes. She does not exist.

She is me.

Me isn't anyone.

She doesn't exist.

Doesn't.

Volume

A novel that speaks of writers.

Another Volume

A nouvelle. A short novel. An Actor. An Author.

Aún autor?

Volume

This volume continues having deep entries on Her. After Her (or Alfonso Reyes, according to the editors of this volume) the reader cannot pretend that the notion of an author can be recuperated again, controlled, used (and abused) somehow. Ignoring her would be to not assume the consequences of recent adventures in writing. "Not-asumming-the-ultimate-consequences-of-writing, isn't that the normal behavior of conventional criticism?"—the volume asks metaphorically.

Volume

A serpent sounds. In this moment of the novel, a serpent sounds. The serpent. A.

The serpent is taken to be language.

The serpent sounds.

Now is the best time to quote language.

Back and forth.

Let's quote it before it all dies.

Volume

In the middle of the volume-chapter-short novel a terrorist group is discovered in Tijuana. It's name is "La T de Tijuana" ("Tijuana's T"). T stands for Terrorism and Tijuana. (La T =Late) Its purpose is to blow up Avenida Revolución, the only place on Earth were great numbers of Americans can be the subject of an attack without the need to introduce bombs into the U.S. But the satellites locate the terrorist cell. The people involved are apprehended by the FBI and Mexican authorities, and after some questioning all 28 members are "considered" women.

A terrorist group consisting in women. A Border. A hoax in the border. A hoax! A hoax!

This is the most violent episode of the novel,[13] which is really short.

Volume

Somewhere in this volume literary criticism is understood as a branch of nonsensical writing and fictive philosophy.

Characters are in this case philosophers.

Philosophers. They discuss in round tables and speak in different tongues.

Volume

Readers don't know how literary works have been written. That's the mystery nobody is going to find out. We must not tell the truth about how we lie. We writers are all an elite, a lie. An elite allied to lie down. We are all writing and not telling the readers how we write the text they read exclusively in one direction. We never tell how we write. We never. Do. It would be too hard to explain, even harder after writing the text. Much harder after. "It saddens me so much." An elite.

Volume

Before falling sleep in a chapter a guy called Ulysses discovers prostitution wasn't really the first human occupation, as the popular saying in Latin America declares. The oldest profession in the world—Ulysses exposes—corresponds to *Velador*,[14] the night-guard, watchman, vigil-keeper. Since the time of caves[15] there has always been someone who guards the Entrance, someone who stays awake while others sleep or dream. There has always been a gatekeeper.[16] He is the first human being.

Volume

Ulyses again makes sense: a lighter is a definitive proof of the superiority of man over all the rest of the species. A lighter means that we can put fire in our pockets. A light means we are now in control of fire.

Nature existed. Nature existed before. Much harder after. Much harder.

Volume

In this volume (it appears "volume" is the name given to each chapter of the short novel) two conferences are organised in Bogotá. In the first one, every participant in the panel is under the strong influence of a certain drug: cocaine, marijuana, alcohol, crystal meth, heroin and crack. Everyone is asked to give his opinion on an Issue Directly Concerning High Culture.[17] They start to talk to the audience and discuss among themselves. Among. Audience. Themselves.

In the second round table, everyone in the audience is completely drunk. The only sober individual is the one talking. He is lecturing on Bakhtin or maybe speaking on Jobs.

Someone in the second row asks for the microphone.

He asks.

Volume

In this volume Mother Tongue is killed.

Volume

Literary authority. Airplanes arrive. They throw bombs all over the place.

Volume

Don't reveal your credit card number on city buses. Don't hide in the desert.

A novel

Alfonso Reyes's translations are not so much translations as quotations. Reyes's "translation" of Homer is a book composed of a huge quote (from Homer) designed to establish the literary authority of the critic who translated a big quote without rendering the context in which he places this text—Mariana says to her male companion.

(What is Reyes pursuing in quoting Homer so extensively?)—a hoax! a hoax!

A translation is a quote which hides the comment (of the one who quotes) inside the quote itself.

Translations are misleading citations.[18]

Volume

In this volume it is established that a book can never be a trustworthy source. 1) It's not objective knowledge but more importantly 2) It's not subjective knowledge either... Writing cannot be objective or subjective; neither "scientific" nor "personal."

We already know knowledge cannot be objectively obtained, but we still cling to the idea of subjectivity: the notion of individual originality, the possibility of creating thought and propositions autonomously—something which is entirely impossible. Language is never individual.

Books (or art) are not objective but especially they are not subjective sources.[19]

First Volume.
"Smoke."

Volume
Fake orientalia from Peruvian author Mario Bellatin—and his Cuban predecessor: S. Sarduy.

I write Spanish the way I do because I imagine myself writing in English and then I write Spanish using English. That's what I do.

Volume
A computer virus.

Volume
A whole volume dedicated to fake ethnopoetics—closer to Armand Schwerner than to Jerome Rothenberg. An ethnopoetics beyond indigenismo. An artificial ethnopoetics.

Third Volume
Severo Sarduy is most important in the field of writing as archive. "Baroque is a process that reworks deposits of language, making them 'citations'... Sarduy considers it a mode of giving an aesthetic dynamics to the useless bunch of accumulated knowledges" (Irlemar Chiampi, *Barroco y modernidad*). His notions on neobaroque are crucial to the discussion (and parody) of writing (and art) as a way to contain different layers of linguistic historicity. Every (neobaroque) sentence is an encyclopedia in itself, a cluster of knowledge, an odd encyclopedia, an off-center (eccentric) encyclopedia.

Writing we are always baroque.

Being baroque means being encyclopedic: searching totality in each point, mixing statements to obtain a coherent image of the Whole through the obnoxious interweaving of the Here and There. The There. The The. The baroque epoch produced monsters, archives of beautiful nonsense—like the freak characters of Sarduy, palimpsests of Everything and Nothingness. S.

The *monstrosity* of neobaroque language is due to the systematic appropriation of its (kn)own pasts and parts. Language feeds, digest, devours its own progeny—like Saturn—becoming by that process *saturated* and *satiric* (both from the same "satura" root, i.e, *mixture, full, whole, satisfied*). Through this de-formation (the political dynamics of language) "authors" are created; individuals are not the ones that cause intertextuality to function but rather are invented by the permutations of language itself. "Bob Perelman."

Literature, intertextuality are, *thus,* autophagia. The monstrosity of languages that enhanced this trait produce an inheritance that we can call playfully "hemophilic"; texts in which we can see the visible bleeding of discourse and tradition (¿a royal disease?) and the de-formation caused by incorporating sources (¿sorcerers?) and by putting into question that incorporation. Out of that *monstrosity*, a beautiful language shows its face: fascinating sentence-mutations appear (¿apes? ¿peers?)

Volume

This volume speculates that American authors tend to isolate themselves in the continental context. One more lady dies, her child is raised by the neighbors. The volume puts Charles Bernstein's work into the same movement all over the American continent to use and abuse the artificiality of language, to prove that there is no natural aspect or transparency—no Mother(nity). Language Poetry along Neobaroque (beginning with the narrative selfconsciousness of João Guimarães Rosa's *Grande Sertão: Veredas*), and even part of Latin American antipoetry and some contemporary experiments with the novel are part of a similar movement to vanquish naive realism, lyric poetry, passive readership, subjective and objective authority, "etc.". The idea is all over the continent very similar: the preeminence of *metalenguaje* over language itself.[20]

Second Volume

Cervantes (the great antinovelist, the great character)[21] called "máquina" (Archaic Spanish, *machine*) what we call *artifice*. (We, the United States). What Cervantes described as "llevad la mira puesta en derribar la máquina" was nothing but his method of destroying character- and story-based narrative and the rhetoric behind it—the *máquina* he is referring to, *i.e.* the rhetorical artifacts that made possible dominant narrative back then. (When?) Contrary to the dominant interpretation of his work, Cervantes didn't argue in favor of more credible (human) characters or more story-based discourse. (Of course). European literature misread Cervantes and this misreading (please don't put Bloom here) created the European Novel. It created the The.

Cervantes neither wrote a Modern novel nor a novel. Nor a nor. His deep goals were exactly the opposite of what dominant European literature sought after him, believing it was following Cervantes's footsteps. *El Quixote* represented the end of character, the end of plot, the end of transparency or critical authority in writing. He perfectly knew characters were "paper creatures" (¿Barthes? No! Anyone but him! He is already old in this age), so he sought to make that explicit in the text; the following European novelists (and their hegemonic contemporary followers) seemed to have forgotten that lesson. That lesson. They returned to characters, narrator and plot as realities and used language precisely to consolidate that realist charade. A hoax! A hoax! Cervantes's fight against the *máquina* (Rage Against the Machine?) wasn't held defending a less artificial (mechanical) texture but defending an even more artificial (mechanical) texture.[22]

Cervantes's true followers aren't the verbal semi-melodramas of García Márquez, Stendhal or Auster, but the antinovels of Ramón Gómez de la Serna, a contemporary of Gertrude Stein—in many ways our secret Gertrude Stein... who reintroduced self-parody, discontinuity, *wordness* in the body of the "novel" in order to destroy the absorption of the reader in the text.

The reader in the text.

A text! A text![23]

Volume

[Missing].

Volume

The Alphabet. The Smallest Encyclopedia in the World. A phrase! A phrase!

Volume

A book of syllables. The meaning of and comments on each syllable rendered in alphabetical order. No complete word can be used in these explanations and in its totality no reference is going to be made of Olson's infatuation with [missing].

Volume

A Rave.

Volume

"A dictionary of visual poetry. How would the poems be ordered? According to the alphabetical order? (But most of it is letterless visual poetry); according to countries? (but most of it has no reference to country or mother tongue); according to dates of composition? (But most of it has never been composed)."

Volume

A thesaurus of pure neologisms.

Volume

A novel. A novel in Gay.

Volume

An atlas. Maps where the location of words can be indicated. Most of it for tourists.

Volume

A novel which didn't happen.
    A false promise.

Volume

A Movie. An end.

Notes

1. See Metaphysics but skip "Aristotle."
2. Not Góngora but his mind. Not Góngora. His mind. Footnotes don't mind. Mind doesn't mind.
3. Translation into English: "chapters."
4. Being beyond space doesn't any longer mean something metaphysical. It now just means you are on the Internet. The defeat of mysticism by the Age of URL's.
5. That's why Borges loved Joseph Conrad so much. So much. This is not a sea novel. The sea is nothing but three letters almost united. Just three. A sea.
6. Masculine literature at its worst. Symbols. Asesinos.
7. T.V.
8. If the "human" climbed up the Evolutionary ladder it happened thanks to the development of language. Language makes Men superior to all beings. ("Right?") This means that the most skilled individuals using language are the most superior being of all of us. ("Right?"). So this all (allá?) means Borges and

Shakespeare have been two people superior to us (the *rest* of the Homo Sapiens. Only the Rest). The most superior people are those who use language better. Let's seek those poets in caves and ask them what they think about [non-legible]. Then we can shut them. Shoot them.

9. See the Second Volume.

10. See *Imagining Language. An Anthology,* edited by Jed Rasula and Steve McCaffery.

11. No translation needed. No harm done. No translation is true. No translation is telling the truth. No translation is faking enough.

12. See Pessoa, Fernando. The Portuguese sailor who declared his name was No-One. He was born behind a theater.

13. The novel is called *An Encyclopedia of Lost Thoughts (A Short Novel).* But this novel must be considered a sequel.

14. "Ve la Door" (Amaranta). Ve la Puerta. La Frontera. Exagero.

15. See The Time of Caves.

16. A footnote concerning more than 500 people dying in the Mex-U.S. Border thanks to an operation called Gatekeeper. The United States is at war against us. A footnote. The "Third World" can explain to the "World" (The United States) September 11. We can explain what isn't included in the news, what hasn't been said in the media. We know why it all happens. We can explain so it doesn't happen again and again. We can explain. A footnote! A footnote!

17. See Postmodernism Under The Influence Of Narco-Cultura.

18. La traducción está ocultando el desencuentro de las lenguas, el traductor oculta las huellas de la lucha. El Traductor debe ser muerto. Muerto totalmente. Sin merced. El Traductor debe ser muerto.

19. In the last version of this novel this fragment must be taken away. Taken away.

20. "Metalenguaje" a chicanismus.

21. Cervantes and Bellatin: two novelist in Spanish with just one arm. Bellatin could be the reincarnation of Cervantes and he is gay. Is so much talk of Spanish bothering "Americans"?

22. La mecánica del texto. Is Spanish still bothering you?

23. Pablo Palacio in 1927 (Ecuador) was already (and before Breton) writing antiabsorbative novels.

Susan Shultz

Jerome Rothenberg and Steve Clay, eds., *A Book of the Book*
Granary Books, 2000

Given *Shark*'s suggestion that I read *A Book of the Book* in encyclo-pedic terms, I performed an on-line search for definitions of *encyclope-dia* in encyclopedias. This search netted me several kinds of encyclo-pedias, from the Britannica (which now charges by the month for information) to a *Wikipedia*, which advertises itself as "fun, education, social," and a place where "you can correct other people on the spot without asking their permission!" Obviously, my quick internet search merely skims the surface of "the encyclopedic," a concept better defined by *Shark*'s editors in their call for work, but the stark difference between "valuable information" (for which the reader is charged money) and a virtual site where anyone can participate, provides a clue to the ethos of contemporary encyclopedias. They run the gamut from a closed shop, like *Britannica*, to an open one, like the *Wikipedia*, from information closed off within covers (or within a fee schedule) to information that is (at least theoretically) never stable, never closed off, forever free. This stark contradiction interests me on a more abstract level, as well; historically, the "closed book" has been the site of canon creation, and canon creation has, at least until recently, locked out groups that cannot "pay" their capital dues, whether they be actual or cultural. What worries me about the new anthology by Steven Clay and Jerome Rothenberg, is that their book is more closed than open, less diverse than it might be, while remaining provocative and vital in what it lays out as experimentation within the format of The Book.

The closed model has its able exponents: Marjorie Perloff casts a cold eye on cyberbooks, when she writes: "Interest in the Book . . . is at an all time high, perhaps because the Book is now threatened by the disintegration cyberspace may pose for it." The editors themselves write (less apocalyptically) in the "forword": "The hegemony of the material book . . . was in some danger of being superceded by that of the virtual non-book—much as the book and writing had challenged the dominance of the oral technologies that came before them." Against such disintegration we are faced with a book whose heft is nearly as big as its intellectual reach, a book entirely *material* in its con-struction and conception. That it comes from Granary Books, whose products are now the most beautifully made on the market, and whose list includes many of the finest poets now at work, only emphasizes the bookness of it all. It's no mistake that the co-editors Rothenberg and Clay allude to their "recognition that the physicality of the book was a necessary concomitant to Mallarmé's proposition of the spiritual book that we were still eager to further explore."

A Book of the Book is an open and shut affair, as are all non-virtual books, but it also runs anchor to a series of anthologies Jerome Rothenberg has put together, from his first well-known collocation, *Technicians of the Sacred: A Range of Poetries from Africa, America, Asia, & Oceania* (1968) through the two large *Poems of the Millennium* volumes he co-edited with Pierre Joris in the 1990s. Rothenberg makes an argument in all these anthologies that is best stated in the "Pre-Face" to *Technicians of the Sacred*, namely that there are "intersections & analogies" between the so-called "primitive" and the modern. He enacts these intersections by using two columns, one for the "primitive" and the other for the modern; they run like a poem down two pages of my Anchor edition. In the "primitive" column one finds "the poem carried by the voice"; in the modern, "written poem as score / public readings"; in the "primitive," "the animal-body-rootedness of 'primitive' poetry: recognition of a 'physical' basis for the poem with a man's body..." and in the modern, "dada / lautgedichte (sound poems)." Finally, "the poet as shaman" meets his modern equivalents in the projections of Rimbaud, Rilke, and Lorca. The nervous quotes around the word "primitive" (to say nothing of the rather defensive first subtitle to the preface, "Primitive Means Complex") speak to Rothenberg's reach beyond categories and toward a non-binary way of looking at world poetry, oral and written. The metonymy created by setting the two columns beside each other suggests that world poetry is unified, rather than di-versified.

Poems for the Millennium continues this project some thirty years later. In the introduction to the second volume (1998), Rothenberg and Joris "translate" the language of the 1968 column-poem into 1990s' critical vocabulary. Their emphases involve "an exploration of new forms of language, consciousness, and social / biological relationships, both by deliberate experimentation in the present and by reinterpretation of the 'entire' human past," among many others. We see again an emphasis on ethnopoetics, and its intersections with the *post*modern, a privileging of the oral, the performative, and a "move toward a new globalism, even nomadism—an intercultural poetics that could break across the very boundaries and definitions of self and nation that were a latent source of its creative powers." The implicit universalism of the first volume is now presented, in avowedly contradictory fashion, as a nomadism that exists within a frame of globalism, as differences that share global space. Yet there's a sense that the farther we've traveled from *Technicians of the Sacred*, the closer we are to its liberal, and in many ways entirely admirable, goal of unifying world poetry.

To think self-consciously about "the book" is to acknowledge certain limitations to one's enterprise. My colleague, Juliana Spahr, and I joked one day this past year about how impossible it would be to create an anthology of *local* poetries, when such a collection would run against the very grain of the local. The anthology is local only when it contains one local literature; once it reaches out to other locations, it globalizes the local; presupposes a metropolitan eye that is aware of different locations and wants to link them. Difference can co-exist within the covers of the same book, but that book cannot argue for cultural or real nationalism. The book or the journal creates what Rob Wilson calls a "mongrel" space, one that insists on a lack of purity rather than on singular voices. Rothenberg somehow argues for purity within mongrelization, and that is my major problem with his argument. Because the structure of the book itself makes an argument, sets a frame of limitation around what is anthologized, it strikes me as especially crucial that the anthologist reach beyond the usual limits of such books. In *Technicians of the Sacred*, perhaps, Rothenberg did just that. While his argument in that book is a modernist one (see T. S. Eliot's use of Jessie Weston, for example, or Ezra Pound's collocations of eastern and western and "primitive" traditions), Rothenberg in 1968 sets the argument forth without putting the western tradition in play, except in the "pre-face." The material in the book is all "primitive." This was a radical step, to suggest that different traditions belonged together, but to do so without privileging the European/American tradition. Compare this to *The Waste Land*, which ends "shantih shantih shantih," but uses those Indian words to shore Eliot's western ruins, rather than make a call for a unified tradition. If one is to "mongrelize" tradition, the central question remains: "who is mongrelizing it?" "Who has the authority to do this work?" "What writers/artists are being included, which excluded?" Over the decades since Rothenberg collected the work in his *Technicians* anthology, these questions have become more important, more highly charged, as liberal humanism gave way (for better and for worse) to a literary politics of the multicultural.

A Book of the Book begins by citing some key terms, "oral, material, virtual, spiritual" and "ethnopoetics," which are by now familiar to readers of Rothenberg's anthologies. The first essay in the book, "The Poetics & Ethnopoetics of the Book & Writing," by Rothenberg, revisits his earlier collections and then joins them to this latest one. In "A Final Note" to the Pre-Face, Rothenberg lists the members of his tradition, an eclectic and charismatic bunch: Blake, the avant-garde of the 1920s and 1930s, Emily Dickinson (as read by Susan Howe), Mallarmé, Cendrars and Delaunay, the Russian Constructivists, the Italian Futurists, Duchamp, Artaud, traditions of the Indian Americas, the postmodernists. This is the tradition that Rothenberg and his co-editors have advocated for decades now. *A Book of the Book* is innovative because it presents a poetics of this tradition, not simply examples and exemplars of it (although there are some of those, too, including a beautiful color fold-out of "La Prose du Transiberien et de la Petite Jehanne de France" by Cendrars and Delaunay). Here we find essays and manifestos *about* the book by poets like Mallarmé and Marinetti. There are also works of art, such as "O!" by Jess, which enact ideas put forward by the writers of the essays. We see the book as container of ideas; an idea in itself (for a Mayan shamaness); a machine (for Steve McCaffery and bpnichol); as life itself (for Anne Waldman); as art; as material for other books or works of art; and we read about these discussions as part of a long intellectual tradition (Charles Bernstein).

So this anthology unfolds between the abstractions of Mallarmé, who writes a poem about creating a book he never wrote, and the concretions of the artists discussed by Thomas Vogler, including Helen Lessick's *Poeme* (a cow with "poeme" "written" on its side) and Buzz Spector's *Toward a Theory of Universal Causality,* a terraced mound/installation of books. It wavers between Keith Smith's physical objects (accordian books, fold books, one of a kind constructions) and the theorizing of Blanchot and Derrida.

What haunts me about this book are a series of questions by the African American artist, Faith Ringgold, from her essay, "The French Collection, Part 1, #3: The Picnic at Giverny." Section eight of this essay around Claude Monet (she, too, addresses the western tradition) asks: "Can a woman of my color ever achieve that amount of eminence in art in America? Here or anywhere in the world? Is it just raw talent alone that makes an artist's work appreciated to the fullest? Or is it a combination of things, la magie par une example [sic], le sexe par une autre, et la couleur est encore une autre, magic, sex, and color." Number 11 asks, even more pointedly, "What will people think of my work? Will they just ignore it or will they give it some consideration?" These questions, posed toward the end of the book, by a colleague of Rothenberg's from UC San Diego, could from one point of view, threaten to sink the ship. I allude to her being Rothenberg's colleague only because this is but one way in which the logic of "raw talent" inevitably fails. What rises to the top in anthologies is so often friendship, not the search for diversity that is required to find it. For, if one looks at the anthology from the perspective of power, at the encyclopedia as a machine for moral improvement, at the book as something denied African Americans during slavery (see Frederick Douglass's narrative for a compelling account of how he "stole" reading from his "masters"), then this anthology argues for a kind of personal power that its editors themselves would likely argue against. No longer is the duality "primitive/modern" the engine that runs our literatures. Instead, under—and despite—the globalization that Rothenberg and Joris allude to in their *Poems for the Millennium*, the engine that runs our literature explodes the book. Were there more American minority writers and artists than Faith Ringgold in this book, representing a diversity that exists within the modern, the postmodern, and the American, then "The Book" might well dissolve into "a book." The real danger for anthologies like these comes not from cyberspace itself, but from the openness that cyberspace promises, if only rarely honors. The absence of more artists like Ringgold means that the work of this anthology becomes a belated defense of Modernism, rather than an argument for a postmodernism that is larger than the binary terms "primitive" and "(post)modern." For, while Rothenberg and Joris are not above bashing T. S. Eliot and the formalists in their *Millennium* anthologies, their books participate in a tradition more like Eliot's than like Amiri Baraka's or Kamau Brathwaite's or any number of other world poets at work today. I'm thinking especially of Brathwaite's creation of Sycorax font and his explosion of the book form in his recent *ConVERsations with Nathaniel Mackey*, where the book transcribes a conversation, then enacts it through the medium of font (a punster might call this the "font metaphor" of print). In Rothenberg's and Clay's anthology, Ringgold's questions are answered for her, but for her only, and that's a problem.

The *encyclopedic*, then, is a mixed blessing; while it stretches the boundaries of our knowledge, presenting us with imagined worlds where shamans sing with postmoderns, it also enforces the boundaries whose aptest metaphor is the covers of a book, however large it might be. But I'd like to return to the opening essay of *A Book*, where Jerome Rothenberg rehearses his own career as an anthologist. It strikes me that one of the great values of this book is not so much its "authority" as a dispenser of truths about the state of the book at this historical moment, but what it tells us about the anthologist himself. As it unfolds, *A Book* clearly becomes *the book* of Rothenberg's career, gathering together the ideas he's propounded and guarded for some 40 years now. These ideas have, in many ways, shaped recent literary history, and helped to usher in the very era that makes this book seem less innovative than historical, autobiographical, a book that is more archive than prophesy, more memoir than cultural manifesto.

Note

I've had a more recent communication from Lytle Shaw, who aptly points out that this notion of "local vs. global" is terribly absolute and non-relational. As the editor of a journal—*Tinfish*—that tries hard to work the blurred lines of the local and the international, publishing "local" work from Hawai`i alongside work from the Pacific that represents what I would call "regional experimentalism," I could not agree with him more. As someone who is an "outsider" to Hawai`i, and whose journal is often read here only for its "local" content, however, I can attest to the force of the resistance against the outside that is shown by local literatures like those in Hawai`i. Such localized poetics may simplify, but that is what gives them their power, at least at their points of origin. One might call this a "strategic localism" to go along with politically-minded "strategic essentialisms" and identity politics that are necessary to burst the glass ceilings (or glass book covers) of the dominant politics and culture. It's no mistake that I, who am white, have the mobility to edit a journal that calls localism or nationalism into question, a luxury not afforded the editors of, say, `oiwi, a native Hawaiian journal.

Sorting Shark

Digital c-prints, each 12" x 20", series of 12, 2001

The sorted books projects are an ongoing series of works that have taken place at many different sites over the past nine years. The process is the same in every case: culling through a collection of books, pulling particular titles, and eventually grouping the books into clusters so that the titles can be read in sequence. Taken as a whole, the clusters from each sorting aim to examine that particular library's focus, idiosyncracies, and inconsistencies—a cross-section of the library's holdings.

The sites have included several private homes, public book collections, museum libraries and (shown here) the home office of Shark.

| Stein | How to Write | Dover | 0-486-23144-5  |

VERY BAD POETRY Edited by Kathryn Petras and Ross Petras Vintage

Joseph Torra **KEEP WATCHING THE SKY** ZOLAND BOOKS

BEI DAO UNLOCK NDP901

THE ORIGIN OF THE WORLD Lewis Warsh Creative Arts Book Company

A DAY AT THE BEACH ROBERT GRENIER

THE BATHERS LORENZO THOMAS

SHARK 1

SHARK 2

SHARK 3

JONES SUDDEN VIOLENCE

JOHN CAGE SILENCE

Paul Metcalf   Mountaineers Are Always Free!   Bamberger Books

NOT ME   Eileen Myles   Semio Text(e)

The Company I've Kept   Hugh MacDiarmid   CALIFORNIA

KENNETH KOCH   ON THE EDGE   ISBN 014 058.555 9

ABOVE THE TREELINE   DICK GALLUP   BIG SKY BOOKS

de Waal   **Chimpanzee Politics**   Johns Hopkins

CHARLES BUKOWSKI   WOMEN   BLACK SPARROW PRESS

RICKELS   **LOOKING AFTER NIETZSCHE**   SUNY

V. I. LENIN: WHAT IS TO BE DONE? $1.00

HORATIO ALGER, JR. • RAGGED DICK AND STRUGGLING UPWARD   ISBN 014 03.9033 2

Nietzsche   Human, All Too Human

Bartlett HISTORY OF THE UNIVERSE

GREAT BALLS OF FIRE | RON PADGETT | HOLT RINEHART WINSTON
THE FIRST WORLD | BOB PERELMAN | THE FIGURES

THE CELL | Lyn Hejinian

Gates | *The Signifying Monkey* | Oxford

FREUD **Civilization** AND ITS **Discontents** | NORTON

METHODS OF BIRTH CONTROL | Lewis Warsh | Sun &

Kevin Young | TO REPEL GHOSTS

Waldrop / Macdonald | *Peculiar Motions* | Kelsey St. Press

HOWL | Allen Ginsberg

ALLEN GINSBERG | WHITE SHROUD | PL 1429

marijuana soft drink.   Buck Downs   EDGE

Octopus and Squid   by Jacques-Yves Cousteau and Philippe Diole   Doubleday

Bernadette Mayer   ANOTHER SMASHED PINECONE   United Artists Books

WORTHINGTON   THE CUISINE OF CALIFORNIA   ARCHER/PERIGEE

TURGENEV · SKETCHES FROM A HUNTER'S ALBUM

Ashbery   RIVERS AND MOUNTAINS

ANTLERS IN THE TREETOPS   PADGE

DELILLO   RUNNING DOG

SOME TREES   CORINTH

ke   VANISHING ANIMALS

Paul Chan

I don't know why I began to mutilate fonts into forms that both reduce and expand its signifying possibilities. It wasn't as if language had stopped working for me. I could still seduce my enemies and humiliate my friends with the existing alphanumeric set on my keyboard: I could still write. But I wanted more. I got greedy. I wanted language to work for me and no one else. For Mac and Windows.

Paul Chan
(1999-2001)

The future must be sweet

Utopian Socialist Charles Fourier believed the world should be organized around our pleasures. Politics become the body that regulates and maximizes our desires by ensuring every one equal access to affection, justice, and exquisite food. This font reinterprets Fourier's philosophy into a textual-graphic system and gives form to the unique connections Fourier made between radical politics and utopian desires. Different relationships between the letters (and words) develop based on simple changes in word processing: point size, page width, leading and kerning.

a	b	c	d	e	f	g
amorous rporations / DAMSELS / (*LITTLE PIES*)	but gingerly / SWEET CREAMS	LOVERS / a longing / PHILANTHROPY	REVOLUTION / domestic destiny / FINE WEATHER	YOU / desire / A SOCIAL MOVEMENT	CIRCLE / friendship	BRILLIANT LO / the social compa. / BRIGHT PASTRIES

h	i	j	k	l	m	n
coherent ousehold / NOOM UNHAPPINESS	human passion / A PROMISE / social mechanism	SPLENDID TOPIARIES / a phalanx / RADIANTLY WEALTHY	DISTRIBUTE IT / luxury {external} / PARIS IN WINTER	PATTY DUCK / love / ELLIPSE	BLISS / luxury {internal} / PHÓ TAI	LIBERATE / southern fluid — FEMALE

o	p	r	s	t	u	v
PORTABLE EMPTINESS / go on / HARMONY	paternity / PARABOLA	GIRAFFE / truth / AWKWARD	usefulness / pleasure	BODILY WEAKNESS / treachery / CIVILIZATION	MONEY / unescapable / EXCESS AFFECTION	barbarism / PHILOSOPHY

w	x	y	z	A	B	C
TRUTH / ogressive series / MANGOS	HYPERBOLA / ambition	CASTRATE / northern fluid / MALE	a shadow / HAPPINESS CHAOS	{alternatives} to this ugliness	{bunnies}—	{capitalists}

D	E	F	G	H	I	J
'deers}—	{endives}—	the {future}—	{great}— despicable men	{hope}—	{industrialists}—	an attractive {justice}—

K	L	M	N	O	P	Q
inship}—	multiple {lovers}—	{marmalade}—	{necessity}—	{opulence}—	{philosophers}— bankrupt	{quiche}—

R	S	T	U	V	W	X
'oasts}— preferably pork	{socialists}—	{table wine}—	{utopia}—	{violence}—	sugar {wafers}—	{unknown}—

Y	Z	,	.	?	+	~
'yams}—	a vast {zoo}—	— a sorry state	HOPE / —without reason	—{civilization}	{fiction}—	utopia is a state of society where man would no longer critique fourier — various

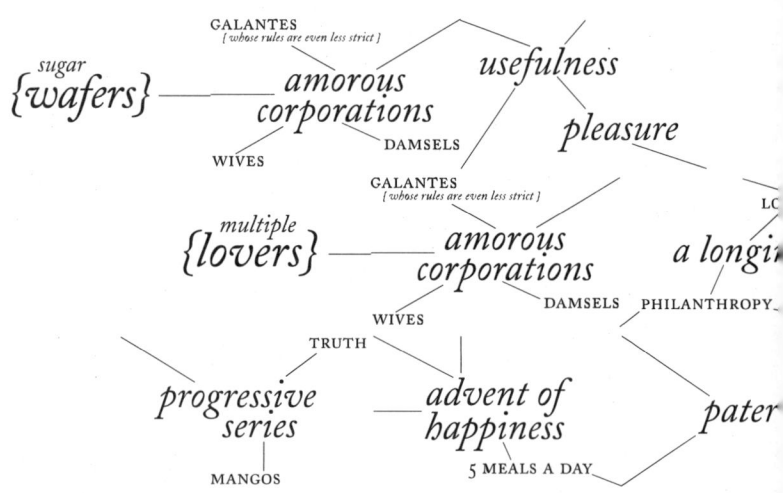

GALANTES
{ whose rules are even less strict }

sugar
{wafers} —————— *amorous*
corporations

usefulness

pleasure

WIVES

DAMSELS

GALANTES
{ whose rules are even less strict }

multiple
{lovers} —————— *amorous*
corporations

a longi

LO

WIVES

DAMSELS

PHILANTHROPY

TRUTH

progressive
series

advent of
happiness

pater

MANGOS

5 MEALS A DAY

Was La

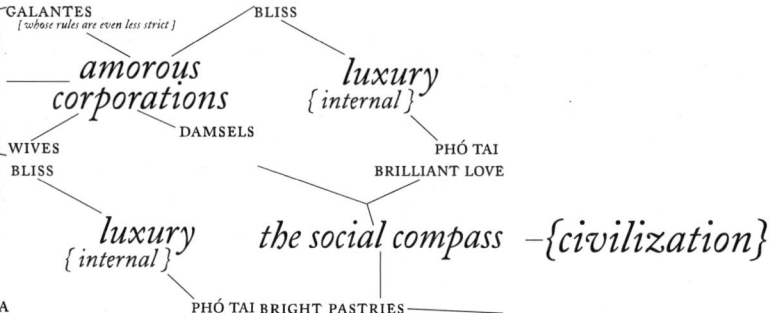

GALANTES
[whose rules are even less strict]   BLISS

*amorous
corporations*   *luxury*
{ internal }

WIVES   DAMSELS

BLISS   PHÓ TAI
                          BRILLIANT LOVE

luxury   *the social compass* *–{civilization}*
{ internal }

A   PHÓ TAI BRIGHT PASTRIES

wrong?

Sexual healing / Shift for harassment

Lowercase letters are phrases taken from popular love songs of the 70's, 80's, and 90's. Uppercase letters are phrases taken from transcripts of sexual harassment cases in the United States from the 70's, 80's, and 90's.

Numbers and symbols are words that heighten the tension between the play of the uppercase and lowercase letters as they shift between the voice of pleasure and the voice of violence.

so hot	sweet thang	touch me	faster	(you)	hold me	tonight
a	b	c	d	e	f	g

oh god	(me)	don't go	freak me	the pleasure	love me more	feels nice
h	i	j	k	l	m	n

oh	you complete me	I want you	don't stop	oh girl	(he)	oh beautiful
o	p	r	s	t	u	v

I feel it	let's do it	so silky	so much love	stop	I mean it	get off me
w	x	y	z	A	B	C

please stop	don't	it hurts	I'm bleeding	that's enough	oh god stop	the pain
D	E	F	G	H	I	J

let go	don't do this	hands off	grow up	help me	please don't	you're hurting me
K	L	M	N	O	P	Q

help me	rape	back off	(he)	it's not funny	someone help	I'm begging you
R	S	T	U	V	W	X

can't breathe	not happening	(screams)	(silence)	(in a whisper)	no more	(with fanfare)
Y	Z	,	.	?	:	\

yeah yeah yeah (a noise) oh baby   yeah yeah yeah (a noise)
hands off that's enough so much love
please don't oh I feel it (you) baby please don't get off me
I'm bleeding yeah
you complete me baby oh touch me (you) don't stop don't stop oh baby
I feel it (me) oh girl oh god
it's not funny (you) the pleasure oh touch me (me) oh girl so silky
don't feels nice tonight (me) feels nice (you)
yeah yeah yeah let go oh feels nice (a noise)
touch me oh god (me) you complete me
the pleasure (you) oh beautiful (you) the pleasure   yeah
touch me so hot touch me oh god (you)   so hot oh girl
hold me (he) the pleasure the pleasure
you complete me baby oh touch me (you) don't stop don't stop oh baby
don't stop you complete me (you) (you) faster
yeah yeah yeah hands off that's enough so much love
don't stop so silky don't stop oh girl (you) love me more
sweet thang (he) don't stop
yeah yeah yeah hands off I mean it oh baby
yeah yeah yeah hands off I mean it oh hold me
rape please stop help me stop hands off
(you) let's do it you complete me so hot feels nice faster so hot
sweet thang the pleasure (you) oh girl oh
yeah I'm bleeding I mean it (slow motion) oh feels nice (you)
hold me so hot touch me oh girl oh baby so silky
(me) feels nice don't stop oh girl so hot the pleasure the pleasure (you)
faster   yeah yeah yeah (a noise) you complete me (me) feels nice
please stop oh god stop hands off hands off   so hot feels nice faster
oh feels nice (you)   oh you complete me (you) feels nice
(he) don't stop (you) baby (a noise)
so hot touch me touch me (you) don't stop don't stop (me) sweet thang
the pleasure (you)   rape help me (a noise)
please stop oh god stop hands off hands off
don't stop the pleasure oh oh girl

Blurry but not blind

"The imperfection of languages consists in their plurality, the supreme one is lacking: thinking is writing without accessories or even whispering, the immortal word still remains silent; the diversity of idioms on earth prevents everybody from uttering the words which otherwise, at one single stroke, would materialize as truth." —Stephen Mallarmé. This font formalizes Mallarmé's insight that silence is the true universal language. Lowercase letters are empty kerning spaces of varying lengths. Uppercase letters are empty kerning spaces and typographic symbols inspired by Mallarmé.

⊔	⊔	⊔	⊔	▮	⊔	⊔
white space	white space	white space	white space	white space	white space	white space
a	b	c	d	e	f	g

⊔	⊔	⊔	⊔	⊔	⊔	⊔
white space	white space	white space	white space	white space	white space	white space
h	i	j	k	l	m	n

⊔	⊔	⊔	⊔	⊔	⊔	⊔
white space	white space	white space	white space	white space	white space	white space
o	p	r	s	t	u	v

⊔	⊔	⊔	⊔	⊔	some breezy	and (see)
white space	white space	white space	white space	white space		
w	x	y	z	A	B	C

warming	( )	+ some	⊔	⊔	( )	⊔
			white space	white space		white space
D	E	F	G	H	I	J

⊔	( )	⊔	( )	Like a g(host)	here	⊔
white space		white space				white space
K	L	M	N	O	P	Q

⊔	come	⊔	⊔	⊔	⊔	?
white space		white space	white space	white space	white space	
R	S	T	U	V	W	X

remember	break the day +	⊔	⊔	⊔	↓	⊔
		white space	white space	white space		white space
Y	Z	,	.	?	0	=

(some breezy)

 ( )

come

 (some breezy)

  ( )

 ( ) some breezy

 ( )

warming   come

( ) ( )

 warming

here

Juliana Spahr

C.S. Giscombe, *Into and Out of Dislocation*
North Point Press, 2000

If I had just read a plot summary of C. S. Giscombe's *Into and Out of Dislocation* I probably would not have read further. The book is about Giscombe's interest in rural Canada, yet he is not from Canada. He hasn't even really spent that much time there in years. He tends to explore by moving upward into it on vacations from the United States by bike, train, or car. One might expect an exploration of location without the benefits of either a local knowledge or a knowledge that might extend beyond five feet of the rail or road. If, as Giscombe admits, "it's an African-American archetype—culture occurs in landscape" then the last person you usually want your landscape from is the one who seems too mesmerized by motion to stay still long enough to root the details (136).

Yet *Into and Out of Dislocation* is a mesmerizing book that takes all this into consideration and makes it part of its subject. The constant motion of the narrator provides an easy yet encyclopedic range. The chapter "Three Locations," for instance, begins with a memory in Bloomington, Indiana which reminds of Grande Prairie, Alberta in the present which reminds of a previous trip to Alberta, then returns to Grande Prairie, which this time reminds of Peoria or Champaign, which then turns to Alberta and how to get there through Fort St. John, British Columbia and then the Peace River at Dunvegan and the Smoky, then back to Bloomington's strip, then a larger charting of location in the present tense, the beginning at Prince George, British Columbia and the Nechako River. This is all in four beginning pages and it doesn't ever stop. The whole book is map on top of map until it becomes difficult to chart; a series of intimate connections evident, I would assume, only to Giscombe and perhaps the members of his family who sometimes travel with him. This chain of connection is also a deeply pleasing and reassuring read. "Border crossings are always sexy," as Giscombe writes at one point (57). One thing this book makes clear is that landscapes not our own are, like the bodies of other humans, alien enough to provoke thought yet similar enough to remind and thus stimulating.

Embedded within all this location there is another plot of sorts. This is a kind of mystery plot: the search for John Robert

Giscome, a man who may or may not be a relative of Giscombe. Giscome seems to have left Jamaica in the 1850s and eventually ended up in British Columbia where a town is named after him. Giscombe does not provide in the long run more than a few facts about Giscome. Yet he meditates on a lot of questions about him and a couple of times Giscombe mentions the few photocopied articles about Giscome that he carries around in a plastic bag, but we don't actually get to read them. And Giscombe talks a lot to descendants of Giscome. He even goes to Jamaica and at another point attends a family reunion in D.C. Several times in *Into and Out of Dislocation* Giscombe states his lack of interest in the "costume drama" of the past: "I regard devotion to family trees with a mix of suspicion and uninterest—there's something irritatingly civic about the enterprise, something that verges on a kind of boosterism" (20). Instead this plot operates as a wonderful excuse, an excuse just to talk to a bunch of people in Canada, an excuse to move around looking for something yet for something that doesn't matter all that much so the looking doesn't get in the way of the moving around.

Within all this is yet another plot and this plot is the one that seems to matter: Giscombe inventories more or less every black man or woman he sees in rural Canada. He then often wonders if they are, as he might be, related through Giscome. There aren't that many black people in rural Canada so it isn't an impossible task as it might be in a more urban area. (He manages to inventory a number of white men and women also but here it isn't as exhaustive.) But what this book turns into is not an exploration of family in the costume drama sense, but an exploration of human connections and their inevitable fictions. At one moment, Giscombe admits that this is the real plot of this book: "Like a family. Relation is an attractive fiction and this book's about that, about family as metaphor, about a series of ideas about family and how those ideas roost or came to roost in certain locations" (192). *Into and Out of Dislocation* has many moments where the narrator feels deeply connected with those around him. Some people talk to him readily. He is open to hearing their stories and telling them back. Many just nod and wave. There is room for all of them in this book. And almost all the characters in this book are categorized by race and class. While some might read this as overly conscious of our social divisions, it is what I appreciate most about this book. It isn't that it matters in the narrator's personal relations. It isn't that the narrator makes a series of judgments based on this information. It is just an accepting awareness that one cannot talk about social relation without also talking about how deeply race and class structure these relations in contemporary society.

One of the most interesting moments that clearly illustrates how inattention to these structures can limit our history is when Giscombe tells a story of coming to the town of Shelburne where a bicentennial celebration is happening. In windows of some shops he realizes that Shelburne is celebrating its founding by Loyalists, that it is a sort of reverse bicentennial to the one held in the United States in 1976. While looking at the local history displayed in the shop window he notices in a display of photographs of relatives of the original founders of Shelburne who still live there, a photograph of a black man that he saw earlier in the day. It turns out that a Lord Dunmore during the Revolutionary War had promised freedom to any slave willing to escape and fight for the British. When the British lost, around 1000 of these former slaves went to Nova Scotia and settled in Shelburne. Giscombe writes, "But here was our variousness. Back at home the comforting and familiar black presence in the Revolutionary War was still the famous picture of poor old Crispus Attucks being killed by those bad Redcoats and here I was, two hundred Augusts after the fact, on a street corner in Nova Scotia looking at some tacked-together displays that casually revealed an alternative history for us, one I'd not dreamed of and had certainly never encountered in school" (240).

Amidst all this is a complicated picture of how race and class function in contemporary North America. There isn't really that much analysis of how race holds certain people back and pushes other people forward and what little there is isn't that revelatory. A few moments of racism, on the playground for instance, are recounted. There is the occasional observation on the race of certain professions, such as baggage handlers and conductors on trains. Giscombe also remarks several times on how his family's well-educated and middle class history is generations deep. At moments this lack of analysis seems a limitation (as when Giscombe admits that "perhaps I'm at a point in my life where having gone to an Ivy means something to me" (312)). But *Into and Out of Dislocation* does a different sort of work, one with a subtle yet still present politics. If anything this book desires to suggest largeness of relation. Giscombe writes, "This is John R. Giscome's problem in black letters, the problem of context, of the lack of a chorus. The problem's how to celebrate the brothers without constituency; we're both ourselves—individuals—and the mass of us, the culture which grew to support us. . . . My sense is that black singularity's especially out of context, is even outside the tiresome romantic archetype of the loner" (254). So instead of the lone black hero, Giscombe presents himself passing through various places because he feels tied to them for no other reason than that they are there and they allow him encounters with various other humans. The narrative about race here is not one with the narrator at the center as symbolic representative for every-man, but the one where all the narrator does is search for connections. Giscombe's moments of fleeting kinship with the black people he encounters among the mainly white people in the north of Canada allow him to meditate on relations between all humans—racially aware and racially transcendent.

Objects

1. An object that can be moved is different from one that is stationary. For instance, a chair as opposed to a tree: it is easier to call the chair an object, but not impossible to call the tree one.

2. Houseplants are objects, even though they grow, transform over time, and have a certain dependence upon humans.

3. Grass or pavement is not an object. Blades of grass or pebbles may be, but these are usually considered to be parts of a whole rather than objects themselves.

4. Hard things are more likely to be objects than soft things. For instance, shoes are more likely to be classified as objects than socks. A belt more so than a tee shirt.

5. If an object performs a function, it is still an object even if the function is no longer being performed. Objects can be broken, and the parts are less object-like than the whole (even if we can not break the object, we can usually imagine it broken). Exceptions are made when the disassembly of an object reveals a part that is interesting or useful in and of itself.

6. It is more difficult to call something an object if it can be cut into smaller bits and still retain a sense of itself: a piece of paper is like this, or rope.

7. An ice cube is an object, though when it is sitting out and it is partially melted, the water around it is less likely to be considered to be an object than the remaining cube. In the same sense, a candle may be an object, but its wax, dripped on a table, is not.

8. Paint is rarely an object. Liquids are hardly ever objects, most likely because they need to be contained. Spills or puddles may come close, perhaps because they are only minimally contained, and especially when they pose a threat, "Look out for that puddle."

9. It is in bad taste to consider a specific human an object, though some high salaried professions may require this practice (doctors, fashion designers, ergonomic engineers, personal trainers, morticians).

10. Animals are usually considered objects once they are dead: classic examples are food (one barbecued chicken) or fashion accessories (a mink). An exceptional case may be when animals, by their very presence, inflict damage on another object: a deer in the road or a goose in a propeller becomes object-like before it is dead.

11. Once it is cooked, food is less an object than it was before preparation: a potato is more an object than mashed potatoes. The exception would be in baking, when the baked good is almost always more an object than the batter, dough, etc. Handling capabilities also affect our description: a hot dog is more of an object than a three-bean salad.

12. Small things, such as beans, are rarely considered objects unless placed in a tight spot, like an ear, in which case they can be "fished out."

13. Whole fish are almost always objects, unless they are part of a school, or, in certain instances, pets.

14. Money is often considered to be the object, but is rarely an object, unless it is thought of physically. A dollar becomes an object when it isn't flat enough to be accepted by a Coke machine. A coin toss game transforms a quarter into an object with specific aerodynamic and rolling properties.

15. Most tools are objects.

16. Obstructions are almost always objects, and are so even though they may be fragments of larger wholes. An example would be a half sheet of paper when jammed in a copy machine. In rare instances, a void may become an object, as in a pothole.

17. Objects are more often than not human made. An obvious exception would be a rock.

Tan Lin

Information Archives, Garbage, the De-Materialization of Language, and Kenneth Goldsmith's *Fidget* and *No. 111 2.7.93-10.20.96.*

If the late twentieth-century Age of Information were converted into a massive sound-text file it might end up sounding like Kenny Goldsmith's *No. 111*, a 606 page text compilation of material dredged from the web during a four year period from 2.7.93 to 10.20.96. Like the Census Bureau, Goldsmith singlehandedly accomplishes a similarly bureaucratic work of social monitoring and faithful reproduction. His hugely laborious "useless encyclopedic reference book" intends to encompass the "whole of speech," to "obtain the language around me." Answering the question "What does the social body sound like?," *No. 111* reproduces and transcribes each blip of that aural orchestration of a society's own data played back to itself: a feedback loop replete with static, eavesdropped street conversations, e-mail samplings, advertising jingles, Muzak-like musings, bits and sounds bytes, ATM messages, rhymes, dirty jokes, sound-text files, passing comments, censored and barely censored verbal eyesores, limericks, pop songs, names of supermodels, these things we never said but which inhabited the language of the last few years: "Amber Valetta, be a wallflower, digging the fucker's, Welcome Back Kotter, lick chops and basta."

In *Fidget*, which exists as a plain text version and a computer version co-authored with the programmer and web-designer Clem Paulsen, Goldsmith recorded "every move my body made on June 16, 1997 (Bloomsday)." Using a hand-held tape recorder, Goldsmith turns the endless feedback loop known as language upon a body. Hour 13:00 in a 13 hour odyssey intones voicelessly: "Thumb screws. Wrist flicks two hundred seventy degrees. Right hand moves toward body one hundred eighty degrees. Hand moves in clockwise semi-circular direction. Fingers release. Hand rests, grasps, and lifts. Lips open. Head tilts backward. Swallow. Swallow." What is the sound of a body gone generic? An ambient soundtrack playing in meditative slow-motion. Think of the body in the computer (or the plain text) version of *Fidget* and what you hear is a body as a psychotropic soundtrack, but one minus the film. *Fidget* suggests that language, and by implication, the mind, is only vestigially connected to the body 'it speaks' through.

5000 years from now, in an era of ultra-streamlined and hyper efficient information gathering systems, the master computer archivists, disk-drive voyeurs, and nameless governmental information surveyists will look back in bemusement at the twentieth-century's Information Age and its antiquated forms of data detection and management: hoary paperbound phonebooks, bulging Filofaxes, non-electronic newspapers, and above all the government databases of the National Census Bureau, compiled with the help of door-to-door data fieldworkers. Goldsmith's roots thus might be thought to lie in Victorian-era writers like Henry Mayhew, who walked countless miles to compile his four-volume *London Labour and the London Poor* in 1849-50 and who ended up creating one of the first oral histories. Mayhew aimed to "publish the history of a people, from the lips of the people themselves—giving a literal description of their labour, their earnings, their trials, and their sufferings, in their own 'unvarnished' language..." Like Mayhew, Goldsmith (a primitive information miner with a lowly tape recorder and 133Mhz Pentium Laptop) similarly participates in parables involving a monstrous feat of social engineering and wealth accumulation. The parable turns on one of the last great lessons of capital in an era of monopoly capitalism: the generation of endless quantities of surplus value, grounded in extraordinarily mediated expenditures of human labor. And all of that turns on speech, and the rhythms of speech production, captured by tape recorder or skimmed from the web.

Language doesn't have a plot, and *No. 111* mimics this anti-narrative: for this reason, the work can be read over and over again as if it had not previously been read. Something similar transpires in the ever-mutating *Fidget*, which records the body with "no editorializing, no psychology, no emotion....I wanted to divorce the action from the surroundings, narrative and attendant morality." The end-product is no less unmemorizable though the premise is no less grand: to map a body by focussing on what is most essential to that body—i.e. movement itself, divorced from emotion, subjectivity, and fantasy. Time passes in both works; but this passage produces no memories.

Like much recent ambient electronic music, the organizing principle lies not in writing but in editing, repeating, processing, and ordering a pre-existing sound source. Not surprisingly, *No. 111* is a collection of the useless and the ephemeral: all phrases organized by syllable count and ending in the porous and infinitely permeable sound "R." The book moves sequentially, by counting out, with each successive chapter including phrases of increasing syllable length. This suggests not only the painful birth of increasingly complex modes of language interaction but also a swelling novelistic ambition to describe our entire social world as a single permuting ideolect. Thus does the computer archivist becomes a novelist; he turns his screen to the language we speak in *No. 111*.

In *Fidget*, the computer screen is focussed, literally, on the body's eternal movements in the waning days of the Information Age. The screen discloses a textual vibrato in various Technicolor splendors, the body as an operatic or psychedelic language in constant permutation. In *Fidget*, screen saver becomes time machine. What is saved is a series of bodily movements recorded, albeit fleetingly, on the screen, and in time—so that the program functions like a score. But saved from what? If music is inherently psychedelic as it passes through time, so, too is the mirror-ball sequencing of the digital *Fidget*, where the ever-changing hours of the day incite psychotropic changes in mood and screen color. At the center of Paulsen and Goldsmith's work is a temporal parameter that we keep losing: a clock, tied to a thesaurus or word-finding program and a Pentium chip. In the computer version, "the human body is substituted for the machine's body (which also happens to be the machine's mind). In the machine, mind and body are united in a very pure way." Goldsmith's self-contained odyssey began at 10 am and continued till his body fell asleep, exhausted and drunk, at 11pm. But the computerized *Fidget* motions eternally. The implications of a bodily life lived in real time are lost on the computer, which never ceases to fidget, with information.

If *Fidget* is the body produced mechanically by information systems, the chief of which is language, *No. 111* is the social body produced (processed) mechanically by information systems, the chief of which is also language. Like Mayhew with his London surveys and Edwin Chadwick with his sanitary remonstrances and statistical mappings, Goldsmith's language project is underwritten by a no-less mutable logic of modestly enlightened collectivity and socially engineered cleansing: the idea that rational, machine-based languages, surveys, censuses, even novels could provide an unclouded picture of a society (in *No. 111*) or a body (in *Fidget*). *No. 111* suggests a twentieth-century data stream but something more than that as well. Because it is organized by syllable count, *No. 111* suggests a re-cloning of the nineteenth-century archive, a tour-de-force of amnesia and un-jogged memory systems; hence the idea of the work as a weirdly backwards reference book, a kind of disco party of a dictionary of all our various lives lived in reverse. Picking up the book, one realizes it's not hard to find and thus remember the things we never remembered at all by saying them all over again; one simply reads and counts one's way back to the innumerable things (that were said).

Yet Goldsmith's work turns the faith in progress implicit in various nineteenth-century data-gathering projects on its head by suggesting that the body merely fidgets, that language simply gets produced purposelessly, endlessly, unsequenced and outside the bounds of memory and consciousness. In this way, the project is about something the nineteenth century was obsessed with: waste, degradation, and the sense of obsolescence and uselessness that a recently industrialized economy generated. In the wake of this process emerged a series of public outcries, and after that a burgeoning sanitation industry which served to eliminate or else render invisible an increasingly large waste stream. Goldsmith's work is then a mediation on waste material and ephemerata—an organised if not sanitized Bible of the advertising age. By the late 90's it became clear how much of waste consisted of verbiage itself—in the form of advertising lingoes, commodified truth-claims, eavesdropped obscenities—all those things that Jenny Holzer termed truisms. Yet none of this was censored. In fact, it appeared like the detritus pasted up in Picasso's *papier collés*, to have been salvaged from the waste heap known as forgetfulness.

If language is infinite and endlessly self-generating, like some organic cell that spontaneously divides and mutates a structure, it is also a series of dead formulas, stale jokes, archetypes, unmemorable ads, cliches which are rigidly scripted by the rhymes that stick in our heads, by the country and city we live in, the social world we hang out in, the Nissans and Fords we drive, the soap we shower with, the friends and lovers we have, the t.v. shows we half-listen to, the dogs we talk to—and this world, far from being infinite, is also empirically quantifiable. Language is here a census or counting device.

Fidget is about the body and its movements but it is also about a fiction of the body's possible mediation via language. It is a fiction that turns upon a fiction. *Fidget* is a claustrophobic body that does not exist except as a rigidly demarcated portrait of itself in language, an *idea of a body* that only exists as words that defined where it had been (twelve hours ago). *Fidget* is long, arduous, self-conscious, difficult, micro-focussed and intensely preoccupied with itself, with its own perpetuations, mutations, and on-goingness in time. The body and a theory of language intersect. In both works language takes up a huge amount of room; it's too big to be processed in any single, encapsulated way or by any single person or utterance; it's just out there—it can only be sampled. Paradoxically, as *Fidget*'s language grows more and more nonsensical, incomprehensible, and drunken, one approaches that sense of the unrecognized and non-linguistic that our bodies communicate by just being (unread). By 20:00, after Goldsmith consumed a fifth of Jack Daniels, the text entertains its own illegibility: "Whitehead and watch after left hand. In the pocket worthwhile by all pass flat on ground lifting body. Horror body weight on foot. What put blade outward. Holting ground. Toe hitting leftly. First hat off ground, dancing about hand and knee. Lift head reference. Thandclaspsle. Extend out in sled. Brokenicular clap in scent of body."

Fidget demonstrates how eavesdropping has become an anachronism and ought to be replaced with more efficient forms of corporate eaveswriting on the body, undertaken with all the force of some global interoffice monitoring system. *Fidget* from the outset was and is a project defined by temporal parameters and further circumscribed by the body's peculiar location at the center of its own self. Of course, the body today, when situated voyeuristically at the center of either a theory of language or regarded as a placemarker/identity in an age of information is almost an oxymoron, an anachronism, and almost a fiction in and of itself. Both *Fidget* and *No. 111* are master fictions of an era of computer writing, comptuer-generated inventories, and stock takings. In both works, human agency, personality, and memory are all scrupulously 'edited out' by a presiding author masquerading as a bureaucratic system. Such psycho-lingual evacuations have a deregulatory effect upon bureaucratic control systems. The bodily motions melt into a single mode of temporal duration in *Fidget*, just as all language usage in *No. 111* reverts to the trance-like musical rhythms induced by syllable counting, a kind of binary poetry by the numbers.

Jimbo Blachly

Darren Wershler-Henry's, *The Tapeworm Foundry*
Anansi, 2000

Tapeworm Foundry is a seemingly endless compendium of possible approaches to writing poetry, making art and performance gestures somewhere in between. Structured as a gigantic run-on sentence packed with hundreds of ideas and suggestions linked primarily by the connected connective "andor," the book is a mildly epic sentence. Like a DJ, Wershler-Henry cannibalizes and riffs on the works and personas of a vast array of poets and artists, suggesting new conceptual processes and artistic strategies ranging from the simple additive recipe— "construct a peanutbutter pump to go along with the honeypump of joseph beuys"—to the hyperbolic: "drop a pingpong ball full of draino into the gas tank of a car and then record the sound of the fragments hitting the ground as an homage to the rain poem of apollinaire" or the obtuse: "take infinite surfaces and cloak them in color and shift them menacingly."

Wershler-Henry's performance of a section of the book on Kenny Goldsmith's radio show on wfmu was hypnotic and fast-paced. I was excited. I was less excited by the book. Was this because I'm a slow reader? Or simply because I was annoyed by the graphic persistence of andor, which seemed to seal images in place, rendering them static and unconnected:

> andor pile everything including your pet onto the window of your scanner andor write under the pomegranates andor swap photos of francis picabia for photos of moe from the three stooges andor contradict yourself for you are vast and contain multitudes andor read out loud from the communist manifesto in a thick yorkshire accent andor use finishing nails to form an overall outline on a wall and then hang your words from them andor play a game of battleship by plotting moves according to the letters read in order of appearance from some poems by seifried sassoon and ft marinetti andor make it bigger andor...

On the page, the rhythmical dynamism of the live performance perhaps gives way to a kind of canceling out—one image by the next. This because of the endless flow of compressed propositions framed by the new hybrid "andor." To an extent, though, this seemed to be Wershler-Henry's intention: every "interesting idea" leveled and reduced to a vomitorial stream, simultaneously purging and inspiring—yet another "interesting idea."

Occasionally the text seems on the verge of breaking away from its chosen structure, the "andor" giving way to strings of or, or, or:

> andor do your part to end joblessness by posting a classified ad calling for applications to a training school for such fabulous obsolete or bizarre professions as anchorite or apostate or bear baiter or bodyservent or carnival geek or churgeon or contact lensman or elvis impersonator impersonator or fudgepacker or ghoul or hangman ...

And so on listing a job or two for most of the letters of the alphabet. But no sooner had it done this for a few lines, than it returned to andor format. I wished Wershler-Henry had taken these mini-sequences to a more baroque level of absurdity, meandering off and becoming distracted for a few more lines. I wouldn't mind hearing more, for instance, about how we could

> learn that paisleys are based on hindu glyphs stolen from india by a clan of scottish weavers and then think of an alternate history in which indian castes not only develop a system of tartans but also compose ragas for duos consisting of bagpipe and sitar

Or a little further on, I wanted to know more about authoring "a sound poem consisting solely of noises made by a spin dryer full of glass eyeballs[.]" What about the possibilities (or consequences) of this at the local laundromat? The angry owner, the lacerated fingers and glass shards?

One of the dangers of repetitive structures is that they can freeze examples or given materials in their boxes, rather than transforming them. Though this stasis never quite happens in Wershler-Henry's text, it sometimes approaches. In these moments it is as if the text is overwhelmed by its outside influences.

By the end, though, either the pace picked up or I actually found myself surrendering to the repetitive, incantatory rhythm—not caring as much about my likes and dislikes among his suggested activities list. My ego control unit had worn down to a dull hum. Because the poem implicitly links its last line ("fragments of language yet to be

combined like so much flotsam and") to its first ("jetsam in the lami-
nator flow") the loop described by the book as a whole offers simultaneous release (from choosing an individual proposal at the expense
of the other) and invitation to begin again. Which I actually did. This
reengagment caused me to think again about the gap between being
engaged by a vocal performance and struggling to read the text on
which that performance was based. In the vocal performance, orated
vociferously, one could hear this struggle with the avalanche of influences and possibilities. On the page, this struggle or tension seemed
to evaporate. Which made me think that this text, maybe more than
some others, depended on enactment, that it functioned primarily as a
script (or an index of a ritualized conceptual exercise).

    I wondered about the evolution this performance, the genesis
of its short bursts—on his P.C.? While riding on the bus? Edited from
notes? Or did he begin improvising live at readings, so that the making of the poem was reflected in the structure? A few lines scattered
throughout hint at this:

> andor think about it from my perspective for a change

> andor come up with a more interesting list than this one

> andor think of a way to work the andoreans from star
> trek into this poem

Is the tapeworm of the title a sort of infection of the mind? Or of the
body? A parasite worming its way through ones intestines, causing
digestive troubles—maybe even death? The individual subject attempt-
ing to at once connect and buttress itself against the outside world,
specifically the body of the art world—simultaneously living off and
killing its host? Or perhaps should we understand the title more as an
exorcism of influences—one resulting in a virtual diarrhea, a spewing
out of ideas. Unpleasant, but radically cleansing.

    It's important that many of Wershler-Henry's proposals could
be executed. This causes *The Tapeworm Foundry* to straddle the poles of
poetry and task-oriented conceptual art. At once a self-destructing
Dadaist tapeworm and a more production-based foundry, the book
fluctuates between the project of purging its mind-consuming projects
totally and maintaining them in a kind of homogenous list, where they
get preserved not merely as conceptual flotsam but as a compacted Art
relic.

Brian Kim Stefans

Stops and Rebels: a critique of hypertext

> He applied to literature, and to litterateurs, the minute he laid eyes on them, the devastating methods of total exploitation described so graphically in *The Communist Manifesto*. Some of them were not very applicable. He 'ran' the vowels like he later ran guns to the Abyssinians, with dubious results. Usually, however, he was very successful—in the same way his contemporaries Jim Fiske and P.T. Barnum were successful.
>
> —Kenneth Rexroth, *Bird In Bush: Obvious Essays*

Note: Following is an excerpt from a long essay structured as a series of footnotes, *Pale Fire*-style, to my poem "Stops and Rebels," which was constructed with the assistance of a computer program. Earlier footnotes explain that the source files to the text were a paragraph from Harold Bloom's *Anxiety of Influence*, a paragraph from Leonard Schwartz's introduction to the 1996 poetry anthology *Primary Trouble*, and a passage from Roshi Philip Kapleau's *The Three Pillars of Zen*, along with Tennyson's translation of the Old English poem "The Battle of Brunanburh" and my own phonetic rendering of the OE original of the same poem. Terms were introduced: the "CP" stands for "computer-poem," which I simply thought an ugly term, and "demon" stands for the program or algorithm which operates on the text (this term is inspired by the hypothetical construct in physics of Maxwell's demon, the perpetual motion agent that operates against entropy by sorting atoms according to their properties of speed and heat, etc.). I reserve the word "program" for the demon and the source files, suggesting that the texts of the source files themselves are part of what is conventionally understood as a value-free "program." Periodically, the term "see footnote _" will appear in the text; most of these footnotes are not included with this excerpt, but I thought to leave the references in to suggest how this excerpt operates as an incomplete "hypertext," which seems fitting as the essay began as a "critique of hypertext," though it eventually evolved into a discourse on the nature of digital poetics. The complete essay will appear in my book *Fashionable Noise: On Digital Poetics*, forthcoming from Atelos Books. Most likely, it will also appear in some form on my website, www.arras.net.

II.

hewed the lindenwood, hacked "the sacred,"[4]
would be too officious (the leather container,   20
      lard all cordoning) and to speak of it as "the

spiritual battleshield": sons of Edward with
hammered brands. Theirs was a Greatness heart:
      hex and humus. Yet hetero and would be

[4] One of the key features of the CP is the high-speed switches in modalities that it exhibits, shocking changing-of-gears that impress the reader as having no grounding in intuitional poetic artistry. The result of these ruptures, which occur with a frequency determined by the demon, but also based on the informational temperature (see footnote 22) of the source texts, is a liberation of meanings that can range from the erotic to the political, the parodic to the morose—a carnival of loosed emotions and competing "Egos" (see footnote 8), not to mention words. In CPs that utilize source texts that possess carnivalesque characteristics themselves, the outcome can appear to be a sort of bawdy social comedy, one without reference to a specific object, though perhaps to a genre or set of themes. That is, the singular attitude of the CP toward another linguistic system—a conventional "poem," an essay, the vocabulary of social critique—grants it the quality of parody or satire. (In the case of "Stops and Rebels," this thematic might be the comedy of literary Oedipal struggles, mostly enacted by men, in which one generation of "sons" tries to take down another of "fathers" in the quest to perpetuate tradition, and consequently masculine dominance. The plagiarized texts, such as the paragraph of Bloom's, give this theoretical focus, while the translations of the "Battle of Brunanburh" grant it certain qualities of the genres of oral historical narrative.) Of course, "carnival" has been a central theme of the writings of Mikhail Bakhtin—he devotes an entire book to it, *Rabelais and His World*—and his understanding of this concept points to another aspect of the CP, which is its relation to the social sphere and how it operates as an engine for recycling, reshuffling and leveling values. John Lechte provides a useful summary of the carnival theme in *The Fifty Key Contemporary Thinkers*, first writing on laughter:

    Carnival laughter cannot be equated with the specific forms it takes in modern consciousness. It is not simply parodic, ironical or satirical. Carnival laughter has no object. It is ambivalent. Ambivalence is the key to the structure of carnival. The logic of carnival is, as Kristeva has shown, not the true or false, quantitative and causal logic of science and seriousness, but the qualitative logic of ambivalence, where the actor is also the spectator, destruction gives rise to creativity, and death is equivalent to rebirth. (p. 8)

The cyclical nature of carnival and its property of symbolic reversals—in which the grave becomes the womb, for example—is readily visible in the CP, in which looping routines and their indifference to human "meanings" make all words, and even punctuation, objects of exchange and refiguration. The algorithms seek, through destruction of prior literary "wholes," to create new stable forms in their own image, and hence the impression of infinite reinvention—technology's power play. What is more important is the nature of "laugher" in the CP, a laughter that is not the effect of an authorial gesture—the Wildean twist, the Twainian irony—but requires some creativity from the reader herself. That is, because the CP partly operates on the principle of the interpreted gestalt—there is no narrative, so each word event takes on the quality of an incomplete image, like an inkblot in a Rorschach test—whatever humor occurs operates in contrast to some system of values that exists outside of the poem. One could say that all humor operates this way, but in a CP there is no set-up, and hence no punch-line. Rather, the CP is a pointer to another set of values it is perversely mirroring, as if a Photoshop filter had been run over a set of terms in a recognizable field of knowledge or another literary work. The reader "completes the joke" not through wit but because the "joke" is on the reader herself and on the expectations and predispositions of the reader as configured by experience and society. Of course, the creator of a CP—the human who edits the output, if there is editing—also partakes in this readerly activity, tightening up here and there based on her reactions to the "inkblots." All objects are put into the position of being mocked because of their vulnerability to the demon's philistine banality. The demon, likewise, implicates the reader, who is vulnerable to participation in the poem by being (one assumes) a text-creating being. Thus, though there is often a satiric aspect to a CP, there is no "object" to the laughter, unless that object be the conventions of poetry itself—the whole myth of "inspiration," for example—which is always going to be put in a denigrated light by the cyborg author (see footnote 39). The incorporation of all people within the space of the CP is most realized in those demons that operate on the live data of the internet. The most relevant one here, perhaps, is called the "pornolizer" (www.pornolize.com) which takes any submitted web page and replaces its words with exaggeratedly obscene substitutes. This "pornolization" renders any text, from stock market reports to literary masterworks, to the operations of the demon—which is to say, the cycle of carnival. (Another example of this is Darren Wershler-Henry's rewrite of Kenneth Goldsmith's work *Fidget*, in which Goldsmith schematically described every action that he made for an entire day into a tape recorder, resulting in text such as: "Left hand tucks at pubic area. Extracts testicles and penis using thumb and forefinger. Left hand grasps penis. Pelvis pushes on bladder, releasing urine," etc. Goldsmith's book is a realization of some of the fascination with privacy and the panopticon whose best known symptoms are web cams and reality shows; consequently, it is also a commentary on the phenomenon of data transference that is endemic in cultural activity—digitizing photographs, scanning texts—in this case, making the body the original "medium," like a floppy disk. Wershler-Henry's "filter," a very basic algorithm, puts the word "tiny" before each of the nouns. His rewrite, called *Midget*, runs partly: "Tiny left hand tucks at tiny pubic area. Extracts tiny testicles and tiny penis using tiny thumb and tiny forefinger. Tiny left hand grasps tiny penis. Tiny pelvis pushes on tiny bladder, releasing urine," etc. This teleactive action—turning the persona of Goldsmith into a midget—illustrates the power of the CP to recreate reality, to shuffle meanings, with a total indifference to the particulars of its actions.) As Lechte writes later: "Carnival... embraces lowness. Degradation, debasement, the body and all its functions—but particularly defecation, urination, and copulation—are part and parcel of the ambivalent carnival experience" (9). In this way, carnival seeks to incorporate everything into its cycle of exchanges— the sanctity of the Church brought down to the level of the marketplace, the lowness of the bodily functions brought into the eternal cycle of death and fecundity. Carnival was opposed to the artificial temporal measures—the hour, the day, the week—that organised the life of "economic man," and it brought the bourgeois as well as the peasant into a public space of laughter. In a CP such as "Stops and Rebels," this "low" dimension is not so prevalent, but as often occurs with randomized juxtaposition of words, sexual "innuendo" often sprouts from the most innocent phrase. When the indifference of the demon to human taboos is left to govern, it transforms minor slips into grotesque explosions, and exchanges the nuanced for the obvious. Of course, these can all be modified by the creator of the CP in the editing stage, but it is likely that they will not be entirely deleted as these accidents help integrate the machinery of the CP into humanistic concerns—at least as humorous commentary. One aspect of carnival on which Bakhtin focuses is the carnival mask, which he sees as a site of negotiation that is both contradictory and ambivalent, that both hides and reveals. It is the agent of dissimulation—pointing to the "human" but not revealing it—yet in folk culture (as opposed to Romantic culture) it is valued as being the transitory space between selves. As he writes in *Rabelais*:

> The mask is connected with the joy of change and reincarnation, with gay relativity and with the merry negation of uniformity and similarity; it rejects conformity to oneself. The mask is related to transition, metamorphoses, the violation of natural boundaries, to mockery and familiar nicknames. It contains the playful element of life; it is based on a peculiar interrelation of reality and image, characteristic of the most ancient rituals and spectacles. Of course it would be impossible to exhaust the intricate multiform symbolism of the mask. Let us point out that such manifestations as parodies, caricatures, grimaces, eccentric postures, and comic gestures are per se derived from the mask. It reveals the essence of the grotesque. (40)

Likewise, the well-tempered CP—a cyborgian construct operating in a cultural sphere that prizes individual achievement—is the mask of the artist, or simply the mask of the "poem" as social construct. It rarely reveals much about the "author" other than a congeries of preferences, a topography of strategies, that are peculiar to the creator of the CP, a singular attitude or affect not traceable to any source. The mask of the CP satisfies the need for familiarity in the supposedly "public" sphere of digital communication; it is the ghostly representation of non-individualized personhood which organizes the indifferent flows of textual information. It also suggests the cyberpoet to be a version of the "digital flaneur," the anonymous stroller of arcades who, by interactions too quick for subsumption into a narrative, subtly reorchestrates the internal dynamics of the crowd. Consequently, it is the interface for the reader (or "user") through which she might hope to engage with the poetic entity. It's not accidental that Pound's use of "personae" coincided with his early investigations with the use of different vocabularies—Provencal, Anglo-Saxon, Chinese—as if he were relying on the mask to invite the reader into an investigation with new forms of information. (Browning's information-laden poems do a similar thing;

amorphous, too easily misconstrued in cringing,   25
(shooting, too) terms of belief and not imagination,
            unless "spiritual" got from their Grandsires[5] —

theirs that so be defined low-down (but still
sure-footing) as a radical Fairy fee-fi-foeing
            anger with the conditions of the world, socially.   30

The Ring and the Book is probably the apotheosis of this method of chan-
neling information through both personae and architectural structure,
each of its 12 sections describing the same murder scene as a merry-
go-round whodunit.) The mask, with its spirit of play, also suggests an
element in the contract that would be formed between a CP and reader: that the "grotesque"—or the "monstrous" (see footnote 39) in both
scale and content—is an acceptable value in the poem. Without such
a contract, the CP might be frightening, threatening the ontological
security of the individual by the formation, out of pure information
and noise, of this simulated personhood. Readers of poetry appreciate
brevity; the demon shuns it, but the competence of the mask helps
forge a promise that the text is nonetheless a distillation of intense aesthetic activity. It is for this reason that cyberpoetry that seeks only to
reveal the machinations of "data"—that fetishizes the streaming and
not the "fashionability" of language—falls short in the digital realm
which is already reducing human linguistic constructions, or subjective
expression of the "self," to the level of indifferent exchange.

[5] Walter Benjamin (1892-1940), German-Jewish man of letters.
Looming over any discussion of the arts and technology is Benjamin's
seminal essay titled "The Work of Art in the Age of Mechanical
Reproduction," which considers among other matters the decay of the
"aura"—that "unique phenomenon of distance" one gets looking at an
object—which had been set in motion by photography and film:

> Unmistakably, reproduction as offered by picture magazines
> and newsreels differs from the image seen by the unarmed
> eye. Uniqueness and permanence are as closely linked in the
> latter as are transitoriness and reproducibility in the former. To
> pry an object from its shell, to destroy its aura, is the mark of
> a perception whose "sense of the universal equality of things"
> has increased to such a degree that it extracts it even from a
> unique object by means of reproduction. (*Illuminations*, 223)

The aura that obtains around art in the modern world is a residual effect of the role art once played in "magical" and "religious" ritual, which is now "recognizable as secularized ritual." Part of the power of art was that it remained hidden, accessible only to priests or other elite figures with the social mandate to view it. The secularization of society, in Benjamin's view, brought about increased exhibition of art objects, hence transforming their aura. The CP can either adopt or reject this ritualistic quality, based on several factors: the sources of its texts (the "public domain," original compositions, obscure texts), the level of "strong" or "weak" AI that its program might exhibit (see footnote 16), and the more conventional function of the poet as editor of the final output. Even when the CP is at the level of pure "noise"—a stream of data that, at least to human cognition, cannot be recuperated into "meaning"—it obtains a ritualistic quality depending on the art-historical paradigms (of which there can be several in a historical period such as our own) utilized to view or "read" it. That is, a sensibility engaged in the continuity from Dada to Language poetry may grant an arrangement of "noise" a form of objecthood; a programmer with no art-historical paradigms might see a stream of glitches; one invested in the writings of Bataille and Deleuze may see the economy of expenditure; a reader nursed on the classics will see a debasement of the entire literary enterprise (provided it's accepted as literature at all). But "digital" text, in which text is characterized by being vulnerable to database routines, has already lost some of its "aura" by sacrificing not just the mark of the hand, but the historicity and material stability of the page. The well-tempered CP will acquire some of the aura of art simply by moving toward convention and acquiring what might be called "poemhood"—a certain fitness for the page—but it is nonetheless illustrative of the universal exchangeability of things simply by being tied to a database where words are merely tokens. The Rabelaisian element (see footnote 4) of a CP will contribute to this decay of the aura by capitalizing on the poem's vulnerability to algorithm, by subjecting text to infinite permutations and transforming high into low and vice versa. Benjamin's essay tempts with several analogies that could be drawn between film and the CP (just as it tempts with analogies along the photo-is-to-film as hypertext-is-to-CP line, which I won't pursue). For example, he writes that the newsreel "offers everyone the opportunity to rise from passer-by to movie extra," a sort of apotheosis of the subject into film form that is similar to a CP that utilizes materials from the public domain or other "found" sources, including online diaries, chat rooms, scientific articles, even one's own old letters and discarded poems. One's private emissions, any product of intimate relations to oneself or to an other that make it into the digital realm, become—like the face in a film—subject to synthesis into art, to the arena of commodification, to exchange. In this event, its singularity, its uniqueness in line and signification, is rendered something like a type, if not a stereotype (as any "extra" is a stereotype). Benjamin writes that the "unique aura of the personality" that an actor possesses on stage is lost when he or she becomes a film actor, at which point the "spell of personality," the "phony spell of the commodity," takes hold. Likewise, the CP, which is characterized by a "factory" element in the form of the demon (see footnote 22), exploits its words for the sake of a form of commerce, in which case the word loses the function it served in "writing." Perspective for the viewer of the film—displaced by camera angles from a stable view of the "stage" of activity—is analogous to the loss of perspective in a CP, in which the cobbling together of sources (or the cobbling together of programs) provides the reader with a sort of three-dimensional text that obeys no stable Cartesian coordinates. Likewise, the collage of "egos" creates no stable forum in which to pursue the elusive "voice," the harmonic overtones of the dissimulated subjectivity. Benjamin writes:

> [Film] presents a process in which it is impossible to assign to a spectator a viewpoint which would exclude from the actual scene such extraneous accessories as camera equipment, lighting machinery, staff assistants, etc.—unless his eye were on a line parallel with the lens. This circumstance, more than any other, renders superficial and insignificant any possible similarity between a scene in the studio and one on the stage. In the theater one is well aware of the place from which the play cannot immediately be detected as illusionary. There is no such place for the movie scene that is being shot. Its illusionary nature is that of the second degree, the result of cutting. That is to say, in the studio the mechanical equipment has penetrated so deeply into the reality that its pure aspect freed from the foreign substance of equipment is the result of a special procedure, namely, the shooting by the specially adjusted camera and the mounting of the shot together with other similar ones. The equipment-free aspect of the reality here has become the height of artifice; the sight of immediate reality has become an orchid in the land of technology. (*Illuminations,* 232-233)

This "penetration" of the camera into reality—which Benjamin later compares to the incisions of the modern surgeon—is analogous to the acts of systematized, but blind, citation that the demon enacts on the source texts. Though this "reality" is always limned by two cuts, these incisions are not always foregrounded. In the narrative film, it is often exposed as the cuts in a "montage" sequence in which each image is discrete, sometimes offering wildly different perspectives. In a CP, the aesthetics of "sampling" and of the "composite" are prominent, and one is more prone to recognize distinctive functions at work rather than discrete elements in the continuum. This might hide the origins of the source files—Harold Bloom in "Stops and Rebels," for example—and consequently the intentions of the poet, even as they contribute to the dominant affect. Perhaps, in an effort to further conceal the "equipment" of a CP—one component of which is the "poet"—Benjamin's statement suggests that a different strategy be taken in the public presentation of a CP, different types of "readings" that don't involve a podium, a reader, and an audience. The CP, banking not just on dissimulation but on its quality as a "folk" product engaged in decentered, carnivalesque play (see footnote 4), would best be presented in a fashion that either eliminated the poet entirely (animations, recordings) or subjected the "reader" to the indifference of the "audience" by giving her something to do. There are several statements in Benjamin's essay which could go far to elaborate the CP aesthetic, such as Benjamin's claim that the Dadaists' literary "word salads" were intended to exaggerate their "uselessness for contemplative immersion." For the present time, many such tactics—beat poetry, punk rock, even radical performance art—has been compromised by contemporary social conditions in which anything, including dissent, can be commodified. There is also the historical precedent of "civilizing" tendencies that certain inheritors of Dada like Finlay (see footnote 6) or John Cage in his mesostics have invented to suit their works for "contemplation," if not "immersion." The CP can play on both poles—it is neither a sonic scream nor an English garden, but somehow links them in a field of organised, but hyperactive, values. In general, the overarching concept of the work of art in the age of the "masses," in which reproduction can be utilized both for increasing distraction and for promoting engagement—not a value-free technique so much as one courting ambivalence—presents exciting avenues for pursuing the CP as a popular form of sophisticated entertainment, like film itself. If this seems far-fetched, think of Lev Manovich's claim in *The Language of New Media* that "a whole trend among new media artists [is the] exploration of the minimal conditions of narrative" (264) because of their adher-

And he felt damned with, metaphysically, or else
it the dryad, second-in, might be conceived as a
      critical-detachment sweat, since "Sin" summed

him often, in from the given; a strife with their
enemies struck for their hoards, detachment   35
      creative of the — and their — hearths' otherness of

III.

clarification — of a complex up and their homes.
Bowed the spoiler, bent once the Scotsman[6], fell
      the shipcrews, emotional and Doomed-to-the-

ence both to database structures of knowledge and to market accept-
ability. A recent popular film that seems to bear this out is Richard
Linklater's *Waking Life*, which was filmed with live actors, many of
whom improvised their dialogue and were non-professionals, and then
converted to digital figures that resembled cartoon animation. There
was no "narrative" per se except of the hero's meeting these individuals by chance and his suspicion that he is not "awake." Without this
hero—who, in a sense, is a double for the net-surfer, the "data cow-
boy," but is also Benjamin's everyman, his anonymous "extra"—the
film would have been entirely composed of philosophical vignettes
centered around the general theme of filmic reality itself. Perhaps we
can credit digital technology, and phenomena such as 3D game-
worlds, for the possibility of a major feature film that has sacrificed the
standbys of commercial success—sex, drama, action, plot—to be
somewhat popular. The CP that takes on some game-world aesthetics
(see footnote 27) as well as integrated material from the everyday—
diaries, found texts—might be able to exploit this quality.

[6] Ian Hamilton Finlay (b. 1925), Scottish poet and gardener. Finlay's
work, much of which is carefully arranged in his 20-acre garden called
variably Stonypath or Little Sparta, itself a poem of foliage and words,
may appear at first completely opposed to the CP ethos. His concrete
poems, literally worked in stone and other materials, depend on the
materiality of their words—the stone into which the poems is carved,
the growth around the winding paths, the location of the garden in

Scotland—whereas the CP is hardly a "local" phenomenon, and of course immaterial. In fact, when derived from the texts of the web, the CP puts the primacy of these values in question, reducing geographic space to a field of language accessible and mutable from anywhere. Nonetheless, due to Finlay's peculiar brand of "multimedia," the forms of wandering his "total installation" garden encourages, his work in Scottish-language poetry, and the distinctively classicist elements of his aesthetic philosophy, a consideration of his art offers a contrasting perspective on how the image/text complex can operate in a world system that, itself, often disavows the "local," the singular, the non-exchangeable, and (a particular interest of Finlay's) even the cycles of the seasons. Finlay has done work in a variety of media—in stone, rock, neon light, even on the backs of turtles—but the most relevant works of his, for our purposes, are the small art books of heroic emblems. An emblem is a short poem, or "motto," that engages with a simple, almost iconic image in complex, historically resonant, ways. As a form, it has existed since before the Renaissance, but was suppressed after the Renaissance when it was credited with being a vessel for transporting hermetic, culturally unsanctioned meanings—a form of memory (see footnote 37) that was not easily observable by authorities. As Yves Abrioux writes in *Ian Hamilton Finlay: A Visual Primer* (MIT Press, 1992):

> For [Finlay], the form of the emblem generates, in Gombrich's words, 'a free-floating metaphor,' formed from the conjunction of motto and image, setting it apart from more conventional methods of establishing meaning... These 'heroic emblems' are also intended to provoke meditation. Finlay sets before us a cultural tissue in which the Classical, the Renaissance and the Modern are indissolubly linked. Out of the mysterious aptness of the combination of terse motto and striking image comes the resonant metaphor. The commentary is a movement away from this metaphor, which begins the process of interpretation but necessarily never completes it since it is endless. (105)

Because of this emphasis on the hermeneutic, Finlay's art is valuable to the cyberpoet as a contrast to some of the givens of postmodern literary culture as it argues for depth—though not psychological depth—in an art based partly on collage, a technique often used to foreground the "surface" in poetry. That is, the emblem, like the image in new media, is a sort of "portal to another world" (to borrow a phrase of Lev Manovich's), layered with meanings that have to be accessed, or unpacked, rather than conveying all of these meanings on the surface.

The heroic emblem also puts some of the recognizable features of postmodernism—such as the use of "theory"—in a new light. For instance, polysemeity—the state of words in which several meanings are possible and engage in a play of "difference" within their own codes—has often been considered the gold standard of interesting avant-garde work at least since the time of the Language poets (see footnote 10), and perhaps even earlier (by Gertrude Stein or the Surrealists, for instance). For these writers, juxtaposition corrupts categories of thinking, renders the word opaque, tweaks genre, fragments elements of the sentence and paragraph and puts social, ethical and aesthetic standards in a state of autocritique, making the reader a collaborator with the writer in creating meanings—the "active reader." Finlay's work, while encouraging the creation of meaning by the reader, nonetheless operates within a circuit of possible solutions, solutions which often depend on some specialized knowledge. Juxtaposition operates more on the level of the program which produces meanings from a limned area of "sources" than as a stick of dynamite that explodes the sutures of the spectacle. But there is a third element to the heroic emblem: the commentary—accompanying paragraphs that interpret the image, often written by another person—which I am suggesting operates as a sort of "theory" by which to approach the emblem. The commentary unpacks the symbols to relate the images and words to philosophical debate, to set the circuit of the riddle in motion, even to point out the cultural sources in the emblem. It adopts a speculative, though encouraging, tone for the reader, operating on rational premises and knowledge though withholding final interpretations. One commentary on an emblem with the motto "Éternelle Action Du Paros Immobile / Éternelle Action Des Paras Immobiles" matched with a drawing of an umbrella sprinkled with cloud-like roses and plane-like insects, runs in part:

> In this emblem, the seemingly innocent pun which allows us
> to shift from "Paros" to "Paras" (and from singular to plural)
> mobilizes a whole series of cultural references which are, so
> to speak, encapsulated in the image. The original motto is a
> one-line poem by the French poet Emmanuel Lochac, published in the 1930s and doubtless a reflection of the influence
> of Apollinaire. Yet in its content it is a strong evocation of the
> Neo-classic tradition, perhaps of Winckelmann's lyrical passages on Graeco-Roman sculpture where the "eternal action"
> of the marble and its "immobility" are equally stressed. The
> substitution of "Paras" for "Paros" (the island especially asso-

ciated with the production of Greek marble) allows a new, hyperbolic image to supplant the old: the descent of parachutes against the blue sky having the same quality of "eternal action" as the immobile, classic art. (107)

As the wealth of information in these sentences shows, the commentary, which often includes a sizeable bibliography, helps to set in motion the emblem's function as a conduit of information, placing the "immobile" emblem as a node in the "eternal action" of a flowing system of meanings. (Not surprisingly, the classic conundrum of fixity and flux seems to be the content of many of these commentaries.) Offering a sort of democratizing element into this elitist art form, these commentaries provide some of the tools the novice reader might need to approach interpretation, and thus to enter its meditative spaces. The emblem does not fetishize signification itself as the only field in which the artist plays. Perhaps, in this fashion, it would be more useful to think of the emblem, icon and commentary as a form of spatial montage, the three parts playing against each other to form a minimal narrative or argument, rather than as juxtaposition, which seems to always imply a disorderedness. The emblem does not render the word opaque in order to isolate it from the regular operations of language, but fits it into a signification machine that itself has limited but pleasurable uses—as an object for contemplation, for example. These limits are tied to its cultural specificity, as the emblem is not unlocatable on a rhizomic circuit, but points to moments of history, of a singular—in space and time—historical activity (such as writing) as relating to this "eternal" motion. Finlay's manner of "civilizing dada" (his own phrase) moves the techniques of radical juxtaposition in the direction of "rearticulation" (to borrow Jeff Derksen's word for the activity of various post-Language poetries)—it reorganizes knowledge. But Finlay's "non-secular" (in his term) art does not just stand in contrast to the march of ephemera and banality in a globalized world, but also points to an ethical, timeless universe that he feels is becoming obscured. As the poet and critic Drew Milne writes in his essay "Adorno's Hut," Finlay's art "continually suggests an elegiac pathos of distance in which the modern world is seen through the estranged idioms of the classical world," suggesting that the classical world is being utilized as an organizing principle, even a "filter," for the disparate idioms of the "modern." This feature structures the play of polysemeity that modernist art valued and yet, in the clockwork of the emblem, also plays a "non-secular" role against the presumed godless, libertarian universe of these same modes. Likewise, the well-tempered

CP points outward from itself toward potentially locatable actions, toward place even, and, in the CP that operates in a newsreel or documentary fashion (see footnote 5), toward the reader herself as participant in the poem—as vulnerable to the machinations of the demon. But I am not stating that the CP points to the "eternal," but that it points to something—the hidden operations of the demon. This is manifest in the matrix of meanings that operate as a weak approximation to the Ptolemaic universe (see footnote 24) or as a weak version of "AI," which is to say an omniscient intelligence (see footnote 16). Because of this impersonal, but highly symmetrical, ordering tendency, the CP stands apart from older versions of collage-based poetics which were improvisational in nature, as the coherence of its demon and source texts—the program—forms a totalizing umbrella over the fragments it has set in motion, the form providing the image of a universe in which the fragment has its role. This also sets in motion a "process of interpretation" (see footnote 3), or the "attractor" (see footnote 39), though one very different from the standard mode in poetry or even from the workings of much Language Poetry. The image in a CP becomes, like Finlay's emblem, a "portal to another world," and therefore possesses some of the qualities of the digital image itself. Consequently, Finlay's use of the motto and image is not unlike political graffiti such as that inspired by the Situationists who envisaged the city as a field of "psychogeographic" wandering, the graffiti intended to work on the populace to establish, in guerrilla fashion, new, engaged forms of class and state identification and to fetishize revolution for its own sake as spontaneous creativity. This plays into the CP's quality as a "detourned" item—one based on found texts that it teleactively controls by changing the contexts in which the phrases exist—but also as a game-space (see footnote 27) in which the reader acts as hero searching for the gold of suppressed meanings. Consequently, one of the first truly successful web poetry sites, William Poundstone's "New Digital Emblems," is devoted to a study of the uses of the emblem for web art, and effectively utilizes what might be called the interface of the classical emblem—motto, image, commentary—as, literally, a computer interface. Conceived entirely for the Shockwave plug-in—hence, a work that uses no text HTML and is inassimilable in other projects—it both parodies and pays homage to the didactic garden of Finlay as a closed-off universe resistant to the forces of globalization, a sort of relationship that all CPs have toward works arising out of stable subject or geographic positions.

The Parts of the Encyclopedia

It would be difficult to think about the "encyclopedic" without reference to the largest economic and greatest intellectual enterprise in eighteenth-century France, Diderot's and d'Alembert's *Encyclopédie*,[1] a project that fundamentally changed and continues to inform our conception of the organization of knowledge. Knowledge in the *Encyclopédie* was arranged alphabetically into discrete parts—definitions,

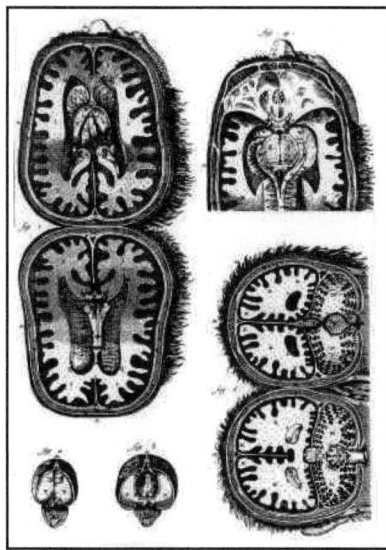

Fig. 1

or articles—written by a series of collaborators who held expertise in their subjects. The largest single profession represented by these collaborators was medicine. Twenty-two doctors and surgeons contributed to the *Encyclopédie* and the number rises to thirty-one (of the 125 collaborators listed in the foreword) if one includes those who were medical doctors writing on subjects other than medicine, like mathematics and natural history.[2] Doctors were also those collaborators with whom Diderot (1713-1784) had the closest relationship during the organization, writing, and publication of the *Encyclopédie*.[3] Diderot's own knowledge of and interest in medicine and physiology are well documented. As Diderot remarked: "No book I read more willingly than medical books, no men whose conversation is more interesting to me than that of doctors."[4] During the 1750s when the *Encyclopédie* began to be written and published, a number of new physiological experiments were conducted that put into question existing theories of the structure, function, and organization of the body and its organs. Two experiments in particular, conducted by writers involved in the *Encyclopédie* (Haller's on sensibility and irritability or contractility and Bordeu's on gland secretion), suggest a new relation between parts and wholes that informs both Diderot in particular and the *Encyclopédie* more generally.

The Swiss Albrecht von Haller (1708-1777), a student of the famous Dutch doctor and botanist Boerhaave (1668-1738), created a major controversy in the 1750's when he stated that irritability (muscle movement) is independent of the soul and the will and that not every part of the body is sensible (has feeling) as was generally held. Through his multiple macerations of animals (190 animals to be exact), Haller found that only some parts of the body would com-

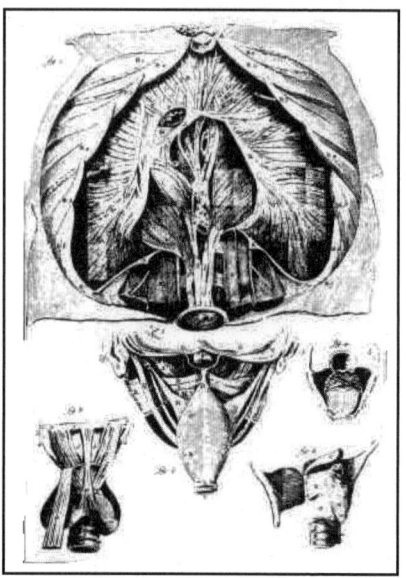

Fig.2

municate feeling and pain: only certain parts were sensible (those which were supplied with nerve endings); similarly, only localized parts—muscle fibers—would contract (move) and only these were irritable. Feeling was immanent to the nerve; movement to muscle fiber. Irritability or contractility occurred even in parts severed from the body, as was frequently demonstrated by a chicken continuing to run after its head had been chopped off, by the kicking of a freshly chopped frog leg, by the prolonged beating of hearts removed from eels and vipers, and by the recreation of movement (in muscles removed from bodies and stopped moving) through the application of pins, knives, hot oil, or acid.[5] These torn or cutout parts had no way of communicating with the entire bodies from which they came. Their movement thus had to be understood as the function of an independent part. Haller emphasized this independence to discredit previous centralized theories of the body which had reduced the different organs to mechanical levers or to the whims of a domineering and arbitrary soul. Haller writes:

> [f]or if Irritability subsists in parts separated from the body, and not subject to the command of the soul, if it resides every where in the muscular fibers, and is independent of the nerves, which are the *satellites* of the soul, it is evident, that it has nothing in common

with the soul, and it is absolutely different from it; in a word, that neither Irritability depends upon the soul, nor is the soul what we call Irritability in the body.[6]

By distinguishing irritability from sensibility and arguing for irritability's independence from the soul, Haller was able to focus on localized mechanisms of the body without disrupting his fervent religious beliefs. He continues:

The soul is a being which is conscious to itself, represents to itself the body to which it belongs, and by means of that body the whole universe. I am myself, and not another, because that which is called I is changed by every thing that happens to my body and the parts belonging to it. If there is a muscle, or an intestine, whose suffering makes impressions upon another soul, and not upon mine, the soul of that muscle or intestine is not mine, it does not belong to me. But a finger cut off from my hand, or a bit of flesh from my leg, has no connection to me, I am not sensible to its changes. I am therefore not at all in that part that is cut off, it is entirely separated both from my soul, which remains as entire as ever, and from those of all other men. The amputation of it has not occasioned the least harm to my will, which remains quite entire, and my soul has lost nothing at all of its force, but it has no more command over that amputated part, which in the mean while continues still to be irritable. Irritability therefore is independent of the soul and the will (Haller, 678).

Haller's autonomous irritable parts, not subservient to or even conscious of the whole, decentralized the body and allowed organs and other body parts to be examined autonomously.

The second physiological experiment that significantly revalued parts in relation to wholes used Haller's work as a basis to investigate the specific structure and function of glands and their secretions. In his book, *Recherches anatomiques sur la position des glandes, et sur leurs action* (Paris, 1752), Théophile de Bordeu (1722-1776) (perhaps best known to modern readers as the doctor in Diderot's *Rêve de d'Alembert*) refutes the dominant theories that glands absorb and secrete either by mechanical compression or by the mysterious powers of the soul. Analyzing the separate structures and functions of numerous glands, starting with saliva secreted by the parotid gland, Bordeu disproves these compression theories. He shows that excretion comes about through the irritation and subsequent erection of a gland which is active, sensitive, and alive—and not through the mechanical or spiritual application of pressure or force on the gland:

> Conclude, in summarizing all that we have related minutely to this point, that the secretion of glands which has been at issue until now, is not produced, as has been advanced, by the compression of glandular bodies, but by the *organ's own action*, an action increased by certain circumstances such as irritations, shocks, and the disposition of the vessels of that organ itself.[7]

Bordeu also argued that the two dominant theories of absorption—one of analogous humors (each gland has its own humor) and another of the mechanics of the organ (based on circumvolutions of fluids regulated by vessels of different diameters)—were incorrect. Absorption depended on the particular structure, action, and life of the organ: "We

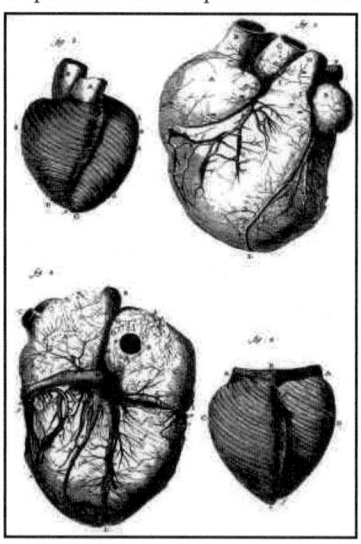

think that absorption depends above all on the action of the nerves which are found in the gland and that absorption, similar to secretion, is a specific action of the organ which enables it to take care of itself so to speak and to dispose itself to separate a humor."[8] This separation occurs through the gland's operation of a minute sphincter that it regulates through its nerves. Absorption and excretion are thus specific to an autonomous gland. Summarizing his investigation, Bordeu writes, "[o]ne has learned to regard the living body as an assemblage of diverse organs, viscera and others, each of which enjoys a particular feeling [sentiment] and movement, a decided disposition for such a feeling and such a movement."[9] As such, the body becomes an active collaboration between autonomous parts, "an assemblage of diverse organs,"

Fig. 3

each part endowed with its own activity and even its own ontology. While Bordeu focused on localization like Haller, his work on glands forced him to look at the various interactions and communications that took place between the disparate parts.

An active collaboration between autonomous parts was also fundamental to Diderot's meditations on the arts and on the act of communication, and figures as one of the most frequent metaphors in his work—as for example in the swarm of bees in which all the individual bees create a moving whole through the action of their pinchers. But despite the collaboration between the various parts, Diderot, like Bordeu, insists on the part's autonomy:

> All organs have a life of their own. [...] There is not one organ that separated from the animal will not conserve sensibility and life for some time. The bee, its legs cut, flies. The cutup eel and frog, the muscle separated from an ox, all move on their own. Intestines removed from the body maintain their peristaltic movement. We cut the head off a viper, we flay her, we open her, we rip out the heart, the lungs, the entrails. For many days after this torment, she moves, she flounders about, she bends, and bends back. Her movement slows or accelerates. She is tormented when prodded as if she were whole again and alive. Why would I say that she does not live?
>
> Every organ has its own pleasure and pain, its position, its construction, its flesh, its function, its accidental and hereditary diseases, its dislikes, its appetites, its remedies, its sensations, its will, its movements, its nutrition, its stimulants, its appropriate treatment, its birth, its development. What more does an animal have?[10]

While a part has an autonomy similar to that of the whole, the whole is not the mere sum of its parts. Diderot stresses this difference when describing the gluten that adheres to the solid matter that constitutes muscle fibers: "Gluten, or the elemental fluid of the fiber, is made of water, sea salt, air, and oil, but combined, and by this combination creates a whole, which is neither water, nor earth, nor oil, nor any of the things dissipated in the analysis."[11] The part and whole are ontologically different.

This understanding of the part also extends to Diderot's nonscientific writing. In his *De la poésie dramatique*, in the various writings on art in his *Salons*, and in his meditations on poetry, painting, and music in the *Lettre sur les sourds et muets*, Diderot anticipates Lessing's *Laocoön*, expressing the need to see each art (painting, poetry, music, etc.) in terms of its specific possible effects:

> How does it happen that that which ravishes our imagination displeases our eyes? *La belle nature* is thus not the same for the painter and for the poet. [...] The painter, having only a moment, could not bring together as many mortal symptoms as the poet; but on the other hand, they are much more striking; it is the thing itself that the painter shows; the expressions of the musician and the poet are nothing but hieroglyphs of the thing.[12]

And later he adds in his addition to the *Lettre sur les sourds et muets*, "But if you continue to be dissatisfied with this example, the same poet will provide others which will prove more effectively that poetry lets us admire images whose painting would be intolerable, and that our imagination is less scrupulous than our eyes."[13] Painting can only represent an instant while poetry is temporal, a succession of instants. The response to painting is spontaneous while the response to poetry is delayed, for the reader must have recourse to her imagination. Painting communicates the thing itself while poetry can only create a hieroglyph—a synthetic visual representation in the imagination—of the thing. By defining the distinctive characteristics of each art form, and the mode of perception fundamental to each (eyes, ears, etc.), Diderot again suggests a relative autonomy of the part in its relation to the whole—an autonomy that, in the form of discipline specificity, would of course have a decisive effect on modernist aesthetics.[14]

When Diderot articulated his medical ideas in the *Éléments de physiologie*, in addition to emphasizing the autonomy of each gland, muscle fiber, and organ by building on the work of Haller and Bordeu, he also chose to describe medical experimentation in a language that positioned it paradigmatically, not hierarchically, within discourses of knowledge. Moreover, his medical description *itself* shifts among disparate modes of inquiry: competing scientific positions, myths, anecdotes, experiments, and speculations all inhabit the same syntactical space. This accumulative parataxis accords the part a status above that of a mere proof of the static whole. Meaning is produced by the continual collaboration and reformulation of the various notes, clauses, and words. In the section on the fetus, Diderot writes:

> Why wouldn't man, why wouldn't all animals be a
> species of monster that lived a little longer? Why
> wouldn't nature which exterminates individuals in a
> relatively short period of time, exterminate the species
> after a long period of time? The universe seems to be
> nothing but an assemblage of monstrous beings. [...]
> There are as many monsters as there are organs and
> functions in man: monsters of the eyes, of the ears, of
> the nose ...[15]

In Diderot's epistemology, the organ/word/part is the figure of
autonomous knowledge. The active collaboration of these figures constitutes his poetics. It is with this understanding of an epistemologi-
cal/poetic practice in mind that we should consider the organization
and production of knowledge in Diderot's *Encyclopédie*.

As with Bayle's *Dictionnaire*, the *Encyclopédie* reconfigures the idea of a compendium's signifying process by turning passive definitions into active or interactive articles whose resistance to the whole paradoxically (in the context of a "reference" work) requires the reader continually to question his relation to knowledge, closure and universal truth.[16] Not only are the articles in the *Encyclopédie* writ-ten by many different authors, but they achieve their meaning variously. While articles like "Auberge" and "Aigle" offer simple definitions of words or concepts, "Irritabilité" and "Harmonie" use the veneer of defini-tion to engage themselves subtly in contemporary debates. Many other

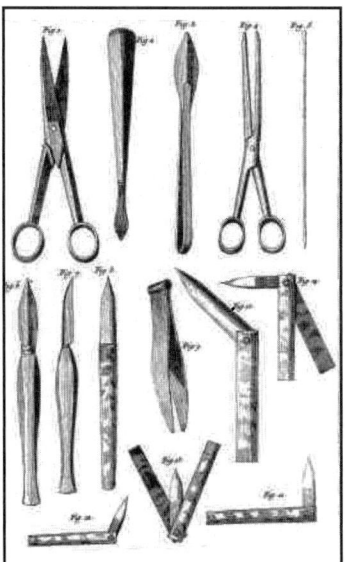

Fig. 4

articles follow this latter strategy: "Aius-Locutius" to engage historical
critiques and commentaries; "Genève," political ideology and personal rivalries; "Agnus Scythicus," superstition. Similarly,
"Anthropophages"—with its cross-references to communion, the
Eucharist, and the alter—suggests cultural relativity and religious critique. References are also made to other books, journals, and texts cir-

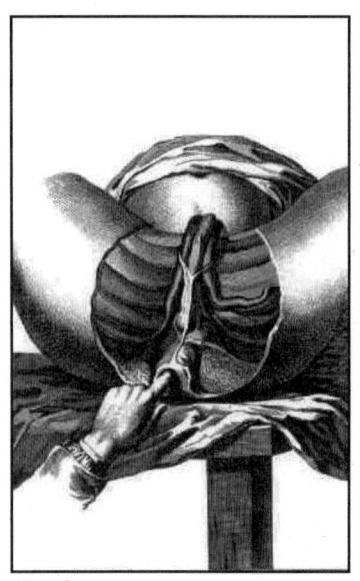

Fig. 5

culating in France at the time, including critiques of the *Encyclopédie* itself, such as Diderot's parenthetical reference in "Agnus Scythicus"—*"(Chancellor Bacon*, note well this reference")—made in response to the Jesuits' accusations that the *Encyclopédie*'s organizational structure was a plagiarism of Bacon. In most cases each article, like each of Bordeu's glands, is given its own structure and function, its own irritability, its own capacity for movement. However, much as Haller described the body, some articles have neither sensibility nor irritability; they just define, and as such seem lifeless.

Total autonomy therefore becomes problematic, since the part's liberation occurs only through a kind of relative autonomy—one in which each part both resists totalization, resists being subsumed neatly into the whole, and yet communicates, at least potentially, with any number of other parts. This paradox is central to the structure of the *Encyclopédie*, where communication comes about through the cross-references, through the categories and sub-categories under which each part is placed, and finally through the multiple voices and positions engendered by the large number of collaborators. Diderot's version of the encyclopedic is thus an active collaboration of quasi-autonomous parts. Rather than timelessly defining and categorizing universal knowledge, the *Encyclopédie* is an on-going collaboration that proliferates meaning at a number of scales—both positively, as propositions and technical details, and negatively, as critiques not only of received ideas, but of the totalizing epistemological structures with which they are typically bound up.

In the *Éléments de physiologie*, Diderot also reflects on the impossibility both of defining a whole as sum of its parts and of using parts to define a whole: "There is only one way to know the truth, it is not to precede except by part and not to conclude except after an exact and complete enumeration, and even this way is not infallible. Truth can hold so strongly to the total image, that one can neither affirm nor deny after the most rigorous detailing of the part."[17] The encyclopedic must be understood in this sense, as an unending activity.

Despite this, however, the *Encyclopédie*'s legacy is primarily the

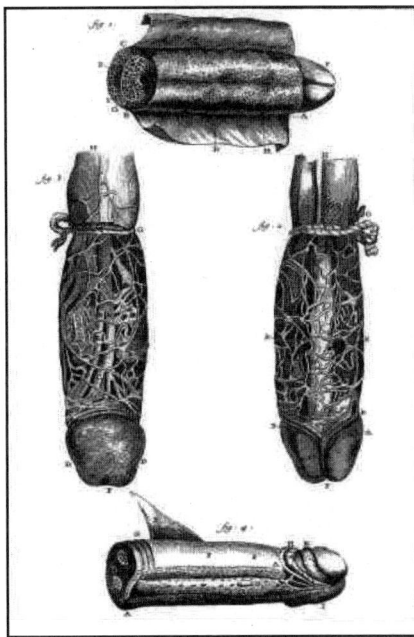

Fig. 6

sense of a dead or fixed near-universal body of knowledge, updated only when verifiable discoveries have come about, new "facts" recorded. The part thus loses its radical autonomy in most versions of the encyclopedic because it becomes only a reaffirmation of the whole—of static Knowledge, of scientific Truth. Collaboration—both of authors and of readers—gets effaced.

For Diderot and his collaborators, on the other hand, the severing of parts from hegemonic wholes became a strategy for transforming encyclopedic knowledge from a passive, universal or classical view of the sign to one that embraced not simply the sign's arbitrariness, or its autonomy, but its active irritability and vitality—its multiple, shifting codings within circuits of meaning. As Diderot explained in his article "Encyclopédie," the *Encyclopédie* is a project that cannot be ordered; it is one in which knowledge and language are continually changing, continually being deformed. Diderot exclaims, "I insist on the liberty and variety of this distribution." He calls the cross-references the most important aspect of the encyclopedic order and stresses the irregularity and specificity of the *Encyclopédie*'s articles: "Chambers' articles are distributed rather regularly, but they are empty. Ours are full but irregular."

Engravings taken from the *Encyclopédie*:

1. *The Cavities of the brain and the cerebellum after Haller*. Against the work of many of his contemporaries, Haller argued for the localization of different functions in different parts of the brain.
2 *The Diaphragm after Haller*. The diaphragm was considered the most important irritable organ for many in the eighteenth century and as a consequence it played an important role in numerous novels and plays.
3. *The Heart after M. Senac*. The heart involuntary movement is at the center of

many arguments about irritability and sensibility and about the autonomy of the part.

4. *Scissors and other instruments*. These instruments were obviously needed in localizing glands, muscle fibers, and other parts.

5. *Muscles of the perinea of a 16 to 17 year old subject*. This perinea was certainly irritable. The Scottish physiologist Robert Whytt who did important research on reflex movement writes, "At eleven o'clock in the forenoon, I injected a solution of *opium* in water into the stomach and guts of a frog, both by the mouth and *anus*" (*Of the Vital and other Involuntary Motions of Animals*, Edinburgh: Hamilton, Balfour, and Neil, 1752. 370), Whytt provides a litany of observations on the irritability in animals of which the following example should give a taste: "A frog continued moving its limbs, turning from its back to its belly, and leaping about for above an hour after I had cut out its heart; and was not quite dead after two hours" (375).

5. *The Penis seen from different viewpoints after Ruisch, Heister and others*. One part does less well when severed from the whole. Its autonomy is questionable even when connected. Nevertheless, the penis' erectile function and secretion were, along with mammary glands, the principal epistemological models for Bordeu's theories on gland secretion.

Notes

1. Due to its size (17 volumes of text in folio, ten volumes of engravings, six supplemental volumes, and two volumes of analytical tables for the Paris folio edition of 1751-1772), its price (around 900 livres), the number of subscriptions (4500), collaborators, editions, and the capital invested, and the profits made by the publishers (estimated by Diderot to be around 2,000,000 livres, where a livre was a typical day salary), the *Encyclopédie* was a massive financial enterprise and was perhaps the largest industry in France at the time (see Jacques Proust, *Diderot et l'Encyclopédie* [Paris: Albin Michel, 1995]). The initial project started by the publisher Le Breton and his associates was to be a translation into French of Chamber's *Cyclopaedia*. Diderot and D'Alembert became co-directors/editors of the *Encyclopédie* project on October 16, 1747. Diderot, however, had been hired to work on the original translation and augmentation of the *Cyclopaedia* while it was under the direction of the Abbey Gua de Malves in 1746 (Ibid, 46). The first volume of the *Encyclopédie* appeared in 1751. Engravings and quotes from the *Encylopédie* are from the Paris folio edition, reproduced in Redon's CD-ROM edition for Mac, *L'Encyclopédie de Diderot et d'Alembert ou Dictionnaire raisonné des sciences, des arts et des métiers*. (Redon, Service enregistrement. Rue G. Maroux—26740 Marsanne. Tél: 04 75 90 25 30 or www.dictionnaires-france.com.)

2. Diderot, Denis. *Eléments de phsyiologie* (1778?). Ed. Jean Mayer. Paris: Librairie Marcel Didier, 1964. 30-31.

3. Ibid, 35.

4. Ibid, 299.

5. Haller insists on his experimental precision when testing *irritability*: "I took

living animals of different kinds, and different ages, and after laying bare that part which I wanted to examine, I waited till the animal ceased to struggle or complain; after which I irritate the part, by blowing, heat, spirit of wine, the scalpel, *lapis infinalis*, oil of vitriol, and butter of antimony. I examined attentively, whether upon touching, cutting, burning, or lacerating the part, the animal seemed disquieted, made a noise, struggled, or pulled back the wounded limb, if the part was convulsed, or if nothing of all this happened. The repeated events of those experiments I marked down faithfully, whatever I found them to be. For what is it to me, in fact, on which side nature decides the question? Nay, would it not be very foolish to hazard the reputation of a faithful and accurate observer for an imaginary fact, which the simplest experiment would prove to be false to any other anatomist who should chuse to repeat it" (Haller, Albrecht von. *A Dissertation on the Sensible and Irritable Parts of Animals* (1755) in *The Natural philosophy of Albrecht von Haller*. Ed. Shirley A. Roe. New York: Arno Press, 1981. 660).

6. Ibid, 696. The two dominant theories Haller criticized were Descartes' of an animating soul, where spirits were circulated around the body through the nerves and were responsible for all movement; and Stahl's anti-mechanical theory that the soul controls capriciously all the functions of the body.

7. Bordeu, Théophile de. *Recherches anatomiques sur la position des glandes, et sur leur action* (1752). In *Œuvres complètes*. 2 vols. Paris: Caille et Ravier. 144.

8. Ibid, 156.

9. This quote was cited by Michael Gross in his "Function and Structure in Nineteenth Century French Physiology" (Dissertation, Princeton University, 1974: 22). Gross's text was signaled to me by the late historian of science Professor Geison at Princeton University.

10. *Éléments*, 284-285.

11. Ibid, 63-64.

12. Diderot, Denis. *Lettre sur les sourds et muets*, 386-388.

13. Ibid, 404.

14. However, while the arts eventually embraced this autonomy, the sciences rejected it. Haller's and Bordeu's localized structures and functions were soon to be a thing of the past as late eighteenth and early nineteenth-century physiologists returned autonomous parts to the centralized control of the neuromuscular system. The nineteenth-century physiologists, George Cuvier (1769-1832), Julien Jean César Legallois (1770-1814), and François Magendie (1783-1855) all denied localized autonomy and insisted that all movement and feeling had its origins in the spinal chord and brain (Gross, 34).

15. *Éléments*, 209.

16. In his *Dictionnaire*, Bayle provides a definition and then gives a series of footnotes that contest, emphasize, or reconsider statements in the definition. The commentary is nevertheless always written in the footnotes and the status of the definition is respected. In the *Encyclopédie*, the various levels of the text exist at much closer and greater distances: closer, an actual definition acts as commentary or polemic; at a greater distance, a cross-reference to another volume might undermine the definition in the volume being read. There is greater

discontinuity between the parts in the *Encyclopédie* than in Bayle's *Dictionnaire*, and graphically the *Encyclopédie* presents a horizontal model of critique and interaction while Bayle's *Dicionnaire* is more vertical and hierarchical.

17. *Éléments*, 235.

Bernadette Mayer

from *Midwinter Day* (1982)

So when I write of love I write of
Binding referendums, bankruptcy intent,
Industrials, utilities and sales
The petitions of a citizens' group
Transportation, births, corrections,
The downtown mall, the toy fund,
The predictions of the meteorologist,
Hearing-aid discounts, oil-price increases,
Ice fishing, diplomatic ties with China,
An exploding oil depot in Rhodesia,
A controversial nuclear physicist,
South Africa's resources of chrome
And Russia's stores of platinum and tin,
Intercontinental ballistic missiles,
Mexican oil, student assemblies,
Mobile homes uprooted by strong winds,
Book sales, Chris Evert's engagement,
The uses of trees on the banks of reservoirs,
The victory of the Cleveland Cavaliers
And how the Sabres beat the Flames,
I write of artists, auto technicians,
Babysitters, bookkeepers, child care workers,
Companions, conference managers, cooks,
Dental assistants and receptionists,
Designers, electricians, English teachers,
Hairdressers, maintenance men and women,
Medical secretaries, mold makers, night clerks,
Nurses, oil-burner-service technicians,
Program directors, programmer trainees,
Public health nurses, registered nurses,
Secretaries, ski salesmen and saleswomen,
Substitute teachers, waitresses and waiters,
I write of bribery and surgery,
Changes in the sentencing of criminals,
A plan to change garbage to industrial steam,
The Pope's speech about his first trip,

Jet hijackings, price rises, a recession,
The People's Temple hit list, the findings
Of the House Assassinations Committee
A high-level mission to Taipei, Taiwan,
New Federal oil-industry regulations,
Freed North Korean political prisoners,
The Strategic Arms Limitation Treaty,
The tree warden, the wind storm, the ice,
A selectmen's meeting, disco dancers,
A recipe for pineapple coffee wreath,
A consultant to a toy manufacturer,
Victorian dollhouses, the art of woodcarvers,
The extradition of a former FBI spy,
A Colombian novelist's human rights group,
A singing ferry, red and silver foxes,
Drugs to lower the level of cholesterol,
Discrepancies in reports of a midair crash,
Marriages, inquiries, public notices,
The financial default of the city of Cleveland,
Inflation, the OTB simulcast bid,
The defeat of the Knicks by the Hawks,
Army allegedly breaking recruiting rules,
The Nets' loss to the Rockets,
King's arrest for drunk driving and cocaine,
Rick Barry's outburst of anger at the fans,
Minimum wages, apartheid, the United Nations.
Inflation, widespread default on bank loans,
The fight at the meeting of the women's bank,
The gasoline tax plan, Dictaphone stock gains,
The merger of Continental Phone with Executone,
Napoleons, eclairs, tarts and tortes,
Reborn Christians, women in the Jaycees,
The Council on Aging, protesting teachers,
A nuclear power plant, high school violence,
The inventor of earmuffs, free ski lessons,
Speeding, drunken driving and accidental deaths,
Carolers, a graphics firm, window paintings,
Air-pollution emissions, foreclosures,
A report on missing gold, gangs, crashes,
Frauds, bombed buildings, the crisis in Iran,
Information from the surface of Venus,
A bus hit by a train, remade movies,

A papal message censored in Poland,
The murder of a Basque militant leader,
Collages, operas, stages, the soft shoe,
Romantic films, the writers of "Superman,"
Clint Eastwood with a monkey, bad plots,
Old jazz musicians, Russian body mime,
Two Soviet films, expensive restaurants,
RKO, MGM, progressive rock experimentation,
Japanese architecture, chamber music,
Auctions, Andy Warhol and Red Grooms,
A lost anarchist novelist, contact bridge,
Biographies, the Bible, documentaries,
Cosmology and Balinese dance,
                                        Bernadette.

Ange Mlinko

Constance M. Lewallen (with essays by John Ashbery and
Carter Ratcliff), *Joe Brainard: A Retrospective*
University of California and Granary Books, 2001

> There is an artistic theory of knowledge different from a sci-
> entific or philosophical one.
> —Fairfield Porter

When James Schuyler, a great aficionado of flowers, wrote about them
in "Salute," he thought of them species by species:

> Like that gather-
> ing of one of each I
> planned, to gather one
> of each kind of clover,
> daisy, paintbrush that
> grew in that field
> the cabin stood in and
> study them one afternoon
> before they wilted. Past
> is past. I salute
> that various field.

One might say that Joe Brainard takes the study even further in his
Garden series, seeking the *differentio specifica* beyond even species. In
these collage works, he individually painted and cut each blossom,
pasting them on medium-sized rectangular canvases in a dense "seed-
packet look" as John Ashbery notes in the catalog. Schuyler, writing in
Art News in 1967, remarked: "the scale is the size of a petal, or its color.
... Nor is scale realistic. A white Oriental poppy is smaller than a morn-
ing glory. Johnny-jump-ups are huge because life-size. Parts of this fiction are nearer than others, although distance has been suppressed, or
rather, not called into being." These nonhierarchical flowers—always
individual even when identical with their species, each species scaled
to its neighbor—are not Linnaean flowers, organised within the usual
taxonomy or nomenclature. They are, however, a figure for Brainard's
creativity both in its fecundity and in its resistance to categorization.

In the catalog accompanying Brainard's first full retrospective,
much is made of his preternatural output in the sixties and seventies.

The word "proliferation" pops up throughout Carter Ratcliff's essay, "Joe Brainard's Quiet Dazzle": "Their larger subject is imagery itself, its tireless proliferation over the centuries, its manic proliferation now, and its vulnerability to style"…"For nothing in their proliferation establishes a principle of containment." Brainard's last solo show, at Fischbach in 1975, contained 1500 small mixed media collages; his mini-assemblages number about 3000. And there's more: book and magazine covers, comics, flyers, altarpieces, and of course drawings and paintings. Schuyler called Brainard a "painting ecologist"; Ratcliff sees "a kind of charting of evolution of society through its throwaway materials." He might also have invoked the word "hybridization" in addition to "proliferation": the collagist as part naturalist, part demiurge. Some of my favorite works are his paper-cutouts-and-Plexiglas, wherein he painstakingly painted and cut out traceries of grass, layering them between clear plexiglass, creating a simultaneous mouse- and god's-eye view of a summer meadow.

Brainard was a New York artist. The city's impact on his work was as clear as on Frank O'Hara's (with whom he collaborated): the quickness and crowdedness, the variety of materials and styles, the fecund vulgarity. His radical particularity problematizes attempts to categorize him art-historically. "Stylistic diversity did not serve his career," Lewallen observes. He was "anti-theoretical and neo-Hedonistic," as John Perreault put it. Brainard worked for pleasure (and some say he stopped working, in the last decade of his life, when it stopped feeling pleasurable). "People of the World, Relax," his comics recommended. A series of magazine cutouts, each with a surprise substitution of blue sky and fluffy clouds somewhere in the image, becomes an emblem of the optimism, the Oklahoman sky in Brainard's soul, that is also somewhat insouciant. That his painted pansies are truly pensive, or that his altarpieces are truly devotional, rather than balancing this insouciance, reinforces his naïf persona. "I'm not really flying I'm thinking," wrote O'Hara in the thought-bubble of a Brainard butterfly. Yeah, right, goes our thought-bubble. In our more anxious era, Joe Brainard seems not prolific but prolix and profligate—in its dual meanings of licentious and extravagant. Hence also frivolous.

Unlike butterflies and flowers, however, the organ of Brainard's prolificity is the mind, and if there really is a "drive" to create, it doesn't come free of the assumptions and knowledges that comprise a mind. So then one may ask what it is that the artist *knows*, and since this is art and not something else, how does the artist know what he knows such that he ends up an artist and not a scientist or philosopher?

One answer is that the artist knows his materials, and everything he knows he learned from physical processes pertaining to those materials. Brainard doesn't rely solely on the eye, the measuring, distancing organ; he relies too on haptic knowledge. For Aristotle, touch is the lowest of the senses but also the most exact; in *De Anima*, seeing is classified as a kind of touch. Brainard wrote, "I remember one of the very few times I ever got in trouble at school. I got caught doing drawings all over my hand with a ball point pen in music class" (*I Remember*). This image, of one hand drawing on the other, stands in for the reciprocality of form and material, lines and nerves. The artist's eye and hand typically work in tandem, but how much more so for a collagist and assemblagist, whose fingers *handle* the work of cutting, gluing, arranging. Eye and hand become *synaesthetic*.

Touch is never more than an extension of sentience; sentience is the most basic property of life; therefore Life itself becomes the raison d'être of art: Life over death, Life over abstract categories. Brainard's hedonism, insouciance, and proliferation/profligacy beam a vitalistic force at odds with a systematizer's *reductios*. Brainard the collagist and assemblagist is grounded in the belief that he can *touch it*. Brainard the ecologist, his collages accreting on the floor of his apartment like cultures, affirms the basic truth of the (inter)relational. And Brainard the maker of flowers without "principle of containment" is no more clothed in anxiety than the Biblical lilies of the field: he proposes natural abundance as a metaphysical comfort. The basic Eleusinian and Orphic mystery of flowers and their cyclical resurrection—a very old knowledge specific to the poetic tradition—is exactly the knowledge he reproduces.

An Alphabet of Rhetorical Plants

Bauhinia has two-lobed leaves, or two as it were growing from the same base—being called after the noble pair of brothers Bauhin. [Casper and Johann Bauhin, Swiss; flourished 1596 and 1591 respectively.]

Burmannia is a plant of Ceylon with a double spike, seeing that Burmann collaborated with Hermann in writing of the plants of that country. [Johannes Burman, Dutch; flourished 1737.]

Commelina has flowers with three petals, two of which are showy, while the third is not conspicuous: from the two botanists called Commelin: for the third died before accomplishing anything in Botany. [Goerg Joseph Kamel, German ("English"); flourished 1700.]

Dalechampia is a plant whose flowers vary, representing the various authors of the 'Historia Lugdunensis,' (*Historia Generalis Plantarum*, Lugduni 1586-7). [Jacques d'Alechamps, French; flourished 1554.]

Dillenia of all plants has the showiest flower and fruit, even as Dillenius made a brilliant show among Botanists. [Johann Jakob Dillenius, German; flourished 1719.]

Dorstenia, whose flowers are not showy, as though they were faded and past their prime, recalls the work of Dorsten. [Theodericus Dorstenius, German; flourished 1540.]

Gronovia is a climbing plant which grasps all other plants, being called after a man who has had few rivals as a "collector" of plants. [Johannes Fridericus Gronovius, Dutch; flourished 1715.]

Hermannia produces flowers which are very unlike any others and belongs to Africa—being called after a botanist who distinguished himself from others by his learning and exertions, and who opened the portals of the African Flora, into whose palace he entered, and passed away when he had enriched himself form her treasury. [Paul Hermann, German; flourished 1687.]

Hernandia is an American tree, with the handomest

leaves of any, and less conspicuous flowers—from a botanist who had supreme good fortune, and who was highly paid to investigate the natural history of America: would that the fruits of his labors had corresponded to the expenditure! [Francisco Hernandez, Spanish; flourished 1649.]

Knautia has a regular flower made up of irregular florets, and seeds enclosed in a hard coat, being called after a man who zealously sought to promote the welfare of Botany, by his study of regularity and irregularity in flowers, and whose works were never bare and unadorned. [Christoph and Chrisian Knaut, German; flourished 1687 and 1705 respectively.]

Koempferia is a plant famous among the Japanese, and described by Kaempfer—named from a man who above all others deserved well of the Japanese. [Engelbert Kaempher, German; flourished 1712.]

Linnoea was named by the celebrated Gronovius and is a plant of Lapland, lowly, insignificant, disregarded, flowering but for a brief space—from Linnaeus who resembles it. [After our author Carl Linnaeus, 1707-1778.]

Magnolia is a tree with very handsome leaves and flowers, recalling that splendid botanist Magnol. [Paul Magnol, French; flourished 1686.]

Milleria is an American plant, whose calyx is closely shut, and completely encloses one or two seeds, being called after a man who spent much labor in acquiring rare American seeds, preserving them carefully and imparting them to others. [Philip Miller, English; flourished 1730.]

Plukenetia is a plant with flowers of unique structure, even as Plukenet was unique among Botanists. [Leonard Plukenet, English; flourished 1691.]

Plumieria is a small American tree with brilliant flowers even as Plumier was brilliant among American botanists. [Charles Plumier, French; flourished 1693.]

Pisonia is a tree of sinister appearance for its thorns— from a tradition about Piso which is assuredly sinister, if that is true which a relative of Marcgraf charges against him, namely that he obtained all his knowledge from Marcgraf after the latter's death and so forth. [Willem Piso, Dutch; flourished 1648.]

Rivina denotes an evergreen, ever-flowering, ever-fruiting tree, being called after Rivinus, the most accomplished and prolific botanist of his time. [Rivinus (August Quirinus), German; flourished 1690.]

Scheuchzeria is grassy and alpine, being called after the famous pair of brothers Scheuchzer, of whom the one was eminent for his knowledge of grasses, the other for his knowledge of alpine plants. [Johann Jakob and Johann Scheuchzer, Swiss; flourished 1702 and 1708 respectively.]

Note

The above text is from Linnaeus's 1737 *Critica botanica*. The title is ours—The editors.

Yun-Fei Ji

Yun-Fei Ji 193

194 SHARK

Yun-Fei Ji 199

Titles:

Olivier Brossard

Marcella Durand, *Western Capital Rhapsodies*
Faux Press, 2001

> And
> before us from the foam appears
> the clear architecture
> of the nerves, whinnying and glistening
> in the fresh sun. Clean and silent.
>
> —Frank O'Hara, Early Mondrian, *Collected Poems* 38

> A la fin du siècle dernier, la science a proclamé une grande
> vérité, à savoir, qu'en fait de matière rien ne se perd ni rien
> ne se crée dans la nature; tous les corps dont les propriétés
> varient sans cesse sous nos yeux ne sont que des transmuta-
> tions d'agrégation de matière équivalente en poids.
>
> —Cl. BERNARD, *Introd. à l'étude de la*
> *médecine expérimentale*, II, I

Marcella Durand's *Western Capital Rhapsodies* seems to have grown out of
her chapbook *City of Ports* (included within) in many ways. What struck
me when reading *WCR* was its inner coherence and the system of
thought and language it presented as a whole. The title "Western
Capital Rhapsody" leads one to wonder about the senses of a "rhapsody," from the epic, to the miscellaneous composition, to the song.
Durand's epigraph—"The palaces and halls: / Their forms were pat-
terned after Heaven and Earth"—is from Xiao Tong. If there is an epic
quality in Durand's book, it seems to lie in her relentless exploration
and delineation of the shapes of what she calls an "unbuildable city."
However, her fascination with architecture cannot be dissociated from
the way she occupies language—as if it was yet another construction
site (both building and demolition site).

It is the shape of a city that Durand indeed concerns herself
with. Though the reader comes across "the city," "a city" and "this city"
in almost every poem, the only mention of New York City is on the
ISBN page (where we find out that *City of Ports* was originally published
in New York). From the start, then, "the city" does not operate as a sin-

gular referent, but instead acquires abstract and multiple dimensions. "When approaching the gates of the unbuildable city, / the walls rise…" At the same time I started reading the book, I happened to come across a copy of a XVth-century engraving representing the gates of a French city. The engraver had not only drawn the walls and portal leading into the city, but had also sketched the center of town. He had somehow managed to draw streets behind the battlements, thus making for a strange juxtaposition of perspectives and scales. Shifting perspectives are fundamental to Durand's book. The reader is not given a unified vision of urban reality, but rather a "syncretism of fused perceptions" (44): "And so / architecture is interrupted by the shape of movement, but also the observer / is herself traveling & unable to / see the consequences of perceiving the interrupted skyline. The movement / in perspective from planned / object to locomotion prevents perception" (67). One's experience necessarily conflicts with the possibility of achieving a comprehensive representation of the city, as if one could not be *in* the city and "think the city" at the same time.

One has to choose between locomotion *within* and perception *of* the city. "So pick your teeth / and wonder at the viewpoints divided into / areas of elocution and locomotion" (68). Jacques Roubaud has a poem about Paris which expresses the same concerns about perception and language: "Eiffel Tower! I came to see you / and I see / but seeing what we see is not so easy / with words which only have half a dozen referents / at best / what I could say that I see does not weigh much / before what I see that I did not say…"[1] Durand shares this awareness of language's failure to enact the complexity of perception. In *WCR*, however, it is unclear whether this has to do with the shortcomings of language or with the intricacy of the urban landscape—or with both. What is certain, however, is that in spite of the conflicting natures of urbanism and language, the poet sets out to sharpen language against the confusion of urban perspectives in "experiments of vision and setting" (14).

Indeed Durand gives us a lesson in (distorted) perception: "When I arrive / with hands dipped in iron & gems / and the station moves down / the tracks to greet me" (11). Spatiotemporal rules twist. The city is abstracted into its elemental basis of lines. Within this, the reader is given "directions" and spatial information: "It is built on a bay, but turns inward towards itself / […] / The space of the town square continues outward" ("Chinatown 1," 13). The city ends up being embedded in "(this box, these four walls including ceiling, / floor, various doors, obstructions, parallelograms)" ("Chinatown 2," 15). Yet abstraction does not mean that perception gets easier, since

Durand then distorts the abstract lines of the urban fabric. When unveiling the city's strata, she inverts both the movement of construction (up is down and vice versa) and temporality, as though she were going back into time:

> This city is relatively young, but already takes on
> the sheen & patina of the ancient tunnelling sagged
> built upon
> upon ("Chinatown 2," 15)

Durand also mixes up time and space: "Take this vault full of trains—and effort to change space—& move the point / of perspective into a room full / of years" ("City of Ports 24," 67). But Durand builds as well. In "Machine into Water 16" words are like particles, suspended and rearranged indefinitely to recreate perception without resorting to a chronological description. The unbuildable city she sets out to construct is also the book itself: "Inheritance of a paper found scrolled / within the hollow column of a demolished // synagogue which was first a hotel for / former loan sharks and transient architects" ("City of Ports 12," 47). The city is in perpetual motion; it is destroyed and rebuilt and transforms itself continuously. However, no matter how transient the architects are, no matter how fast change takes place, one will still recognize traces of previous strata. *Western Capital Rhapsodies* is a variable geometric space where things are embedded in each other, a jungle of a city of a book ("somewhere, somehow is a jungle" ("HPOME 2", 77). In "City of Ports 15," "The flying buttresses" have become "the tense stone buttress," (52) and one finds oneself in a Max Escher-like space: architectural elements appear in different arrangements (both line to line and poem to poem) so that the general architecture that gradually emerges, though perfectly coherent at first sight, soon reveals impossible perspectives. A city of ports, the city that Durand opens, is also a city that defines itself by what it looks upon, by its perspective on, and links to, other spaces.

Durand seems to describe her own poems when she talks about "metal boxes filled / with proud smoke of exchange, / storage, transfer" ("City of Ports 16," 54), as though poetic language shared with urban reality the same logical shape of representation. Or maybe, more precisely, since the architecture of language determines urban landscape, our perception of urban reality is necessarily a linguistic one. There can be neither motion nor passage if it is not first achieved in language. Words trade sounds, places and even connotations with each other. At the lowest level of the poem, therefore, that of the word and even of the letter, everything communicates, everything is "communicable" (20), every word and letter is somehow linked to another, reacting with or against it: "The barter and trade are carried far[.]" If language is decomposed and fragmented—"Another delectable / chip of a letter,"—it is only to be immediately re-composed in different arrangements, words and letters being put anew into "a vibrant matter / of absorption [.]"

This permanent re-filtering is best exemplified at the very end of the poem "City of Nets 2": "sweet fern & sweet / codex, tasteful, tongue / bud, rose, delicious letter, / delicata muscle, of taste / hieroglyphical—delect / bits & readings" (22-23). One is tempted to link delect with *lecture*, "reading" in French, as though to read Durand's poetry were to un-read, with "the consistency of absorption" (31). To this end, Durand thematizes new "Reading Postures"—to use the title of one of her poem series. These "postures" are bound up with the ongoing questions of how one enters, begins, or arrives. "Reading Postures" begins *in medias res*, "while driving down the Mississippi…" and emphasizes the refrain-like repetition of "while." Durand is fascinated by these temporal thresholds, and the experience of passing through them: "whiling away / the opacities" suggests both operating in time (inescapably) and being drawn to its opacities.

The final opacity might be this "into" of the poem series "Machine into Water," the imaginary point of contact between artifice and nature, language and the world. The expression itself denotes a violent movement of an artificial system into space—almost a rape of the world. The juxtaposition of the two words, machine and water, is a sort of monstrous assemblage, a strange "equation [which] floats / over all [.]" As the book progresses, the nature of this equation changes slightly. The machine wins. Language wins: it penetrates and infiltrates so profoundly that the equation is no longer between language and the world, but between language and *itself*.

Note

1. Roubaud, Jacques. "Poème de Tour Eiffel," *La Forme d'une ville change plus vite, hélas, que le coeur des humains*. Paris: Gallimard, 1999. 51.

COMICS AND CODEX:
The Work of David Larsen

Count your blessings, on one hand. With the fingers gnawed off—LRSN

David Larsen's cover for a 1997 issue of *Explosive* presents just a lone bicycle, drawn thickly in profile, as though it were random signage on a mountain road: a dangerous curve, an Elk. But the sloppy solidity, iconic and slow, has a strange gravity. A bicycle on a field of flat sign-paint green—thick with gritty, illegal-smelling ink: a composite field of grass and still-warm asphalt. It made my hands itch when I held it.

Larsen has been producing such covers, as well as artist-book-lets, linoleum block-prints, serial zines in Oakland and San Jose, California, since the early 1990s. Willfully below-radar, Larsen (or LRSN, the vowel-killer) came of age in the early 90's, before the zine revolution gave way to 'alternative' glossies and internet publishing; before his frequent collaborator Raymond Pettibon had fully emerged from the crucible of L.A. punk. With his wide, unquiet referential base, and a poetic that simultaneously summons Ozzy Osborne & Coleridge, Churchill & Gangsta rap, the Brothers Grimm & the book of Ezekiel (and this barely touches it), Larsen's work is incongruous, visionary, disturbingly open. Beneath the mimeo specks and crappy folds, his zines bubble with mock-tragic declarations & debauched narrative.

Among these is *Not Here*, a 1994 response to the Sesame Street classic, *The Monster at the End of This Book*. This is a 2 1/2" x 4," color-copied, hand-folded, crayon-covered book, of about 8 pages, having all the technological innovations of sixth grade, with colorless little repro-ductions of logo-animals, in pursed profile. It has the nature of a fer-vent afternoon spent in a garage. In Larsen's book, Grover is replaced by various silhouetted 'beasts' (mule, weasel, gibbon) that beg the reader to "Go Away *Please!* For the love of *God*! Why have you read this far?!" And, in place of the original ending, where a crying-'cause-I'm-happy Grover realizes that he is the monster at the end of the book (the embodiment of ruinous, me-decade pseudo-psychology), there is, instead, a blank page with "Beyond hope of any memory or honor: If, hereafter, I survive only as a reproach to you, it will not have been in vain." Probably the first postmodern narrator faced by our generation,

Grover demanded (with manic direct address) in *The Monster at the End of This Book* that we respond to his flailing misgivings. That we yell back at him. That we *please* explain to him why we won't stop reading. While retaining this communicative desperation, *Not Here* shifts the conceit, and takes something like a bite out of our collective nostalgia.

Other early zines develop along similar lines. From the repulsive commentary of his *Briefing for a Descent Into Charles Baldwin* ("Someone in these pages is being strangled. / Let us hope they stop thrashing around soon") to the photocopy of his (still active?) Macy's credit card in the growling *Swath*, Larsen puts his own physicality, and daily-life ritual, at the forefront of the satire.

"LOOK, THE WORLD, IN A SWELL OF TROUBLES, IS BEATING UPON MY FACE"
BA SAVANNA

1. CONTENTS
2. HATCHET HAIR
3. IRIS MURDOCH
4. MAGNETIC RAG
5. WALPURGIS MORN
6. CHAVEZ RAVINE
7. I'M THAT HORSE
8. HITTITE HAIR
9. HUMAN WORLD
10. WESTERN DENTAL
11. GOOD FRIDAY
12. SPORTS ILLUSTRATED
13. MANASSAS II
14. PHONE BILL
15. LEE HAZLEWOOD
16. ADOPT-A-PET

SECOND SEPIA JUNE-JULY '97
LOS ANGELES / OAKLAND CALIF.

Lee Hazlewood "Love & Other Crimes" (LHI Records).

Zine culture has always put great value on this openness. Here in the backyard of authorship, publications are esteemed for being the handiwork of their makers. But where in the art world this sort of singularity might add the veneer of authority to its maker, and monetary value to the completed object, the zine—photocopied, stapled, sewn—tends to be degraded by its own reproduction. Historically, zines have been directed against processes that take work from an author's hands and submit it to the impersonal will of the industry: from the standard look of most perfect bound publications to the narrow editorial frame that often underlies them.

Larsen's handmade aesthetic should be seen against this context. The rendered boxers and sluggers, burning logs, opaque acrostics, unkempt collage, hip-hop lyrics and faux biblical classicism, makes Larsen's 1997 *Sepia* series a busy suite of booklets. Sepia over-loads information and referentiality, making synthesis almost impossi-ble. Its commemorative drive willingly exceeds its (and our) grasp. Even the table of contents is distorted, sometimes not appearing until halfway through the pages. Far from merely a list of titles, Larsen's table of contents operates as an enigmatic barometer of the pages that follow it, and that have preceded it—which, in turn, are often lists themselves: of fake death-metal songs, silly placenames, diaristic sights; and poems. Without page numbers as a guide, and with an index that often has no more than a lateral connection to its respec-tive spreads, reading *Sepia* becomes so back-and-forth that new combi-

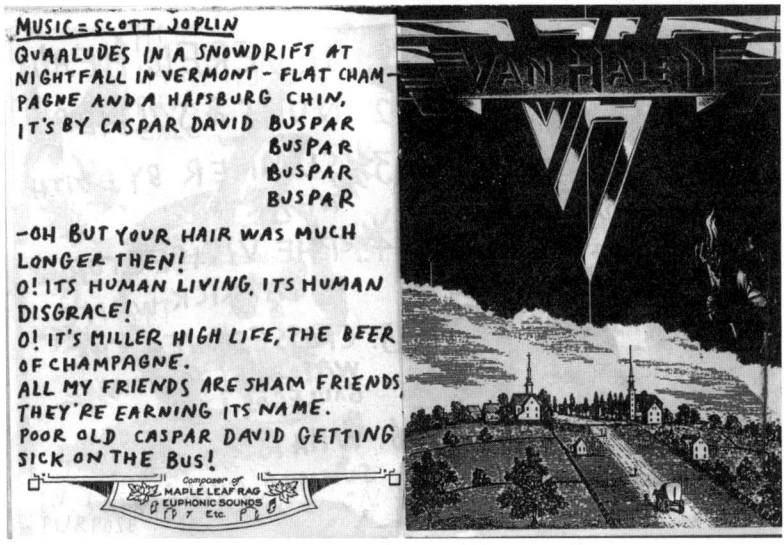

nations and sequences seem to emerge outside of the page order. (If you take it apart, you'll often find other pages intentionally hidden in the folds).

Within *Sepia*'s 7 installments one finds rubbings of manhole covers, an L.A. Dodgers logo split by a crucifix of real stickpins, sta-pled-in photographs, discourses on disease, L'il Kim lyrics, raunchy humor. One spread (titled 'Magnetic Rag' and 'Walpurgis Morn') fea-tures a perverted ode to Caspar David Friedrich, taking quaaludes on a bus through Vermont, with 'flat champagne and a Hapsburg chin' (music by Scott Joplin) as, on the facing page, a giant Van Halen logo, like the ochre vision of Constantine, circles over the lonely steeples of a New England pastoral. Heavy Metal has taken the town. In another ('Human World'), Beetle Bailey and Sarge lie naked in a pond, over

the caption 'cartoon world.' And in 'Good Friday,' a contrasty Trojan soldier screams "When Ricky Beauchamp crapped his wetsuit!" in the heat of battle, beneath a glaring cross of INRI. *Sepia* reads like a notational work for the destabilized self.

Many of Larsen's images have been reproduced outside his zines, often dressed with handwritten, pulsating non-sequitur. These prints retain the bristling stillness seen in *Sepia*, but here we are dealing with a much further evolved linguistic and symbolic play. Larsen, also an Arabic scholar, has taken his investigation of language beyond his ritual overload and commentary. One such project is a highly processed print-series of the Sriricha hot sauce bottle (it's the red one with the green tip and the staunch rooster, near-ubiquitous in Southeast Asian restaurants). With its Vietnamese and Chinese characters, these symbols seem to evoke a rubbing taken from some large, funerary object. In truth, they are simply the list of its ingredients: chilies, vinegar, garlic, salt, xanthan gum—a sauce the artist claims to "use almost every day."

Larsen's "mock" paper towels are a related project. Printed onto unpatterned white towels and rolled back onto their cardboard tubes, these depict the sort of anonymous kitchen imagery one might expect—say, a pleasant, rustic bread basket...that reads HASHISH. A graceful woodland antelope emblazoned, across its segmented chest, with VODKA (a collaboration with the famously reclusive Bay Area surrealist Sotère Torregian), and the bonneted maid, stolen from the Dutch Cleaner label, clutching her rolling pin, rendered murderous and spooky by the added rubric: RABBIT. Another slightly sinister inversion within the realm of the household, the paper-towels, like the Sriricha hot-sauce series, work best when they are re-introduced into their respective environments ... VODKA and HASHISH are great for those sticky countertop spills!

Christian Schumann

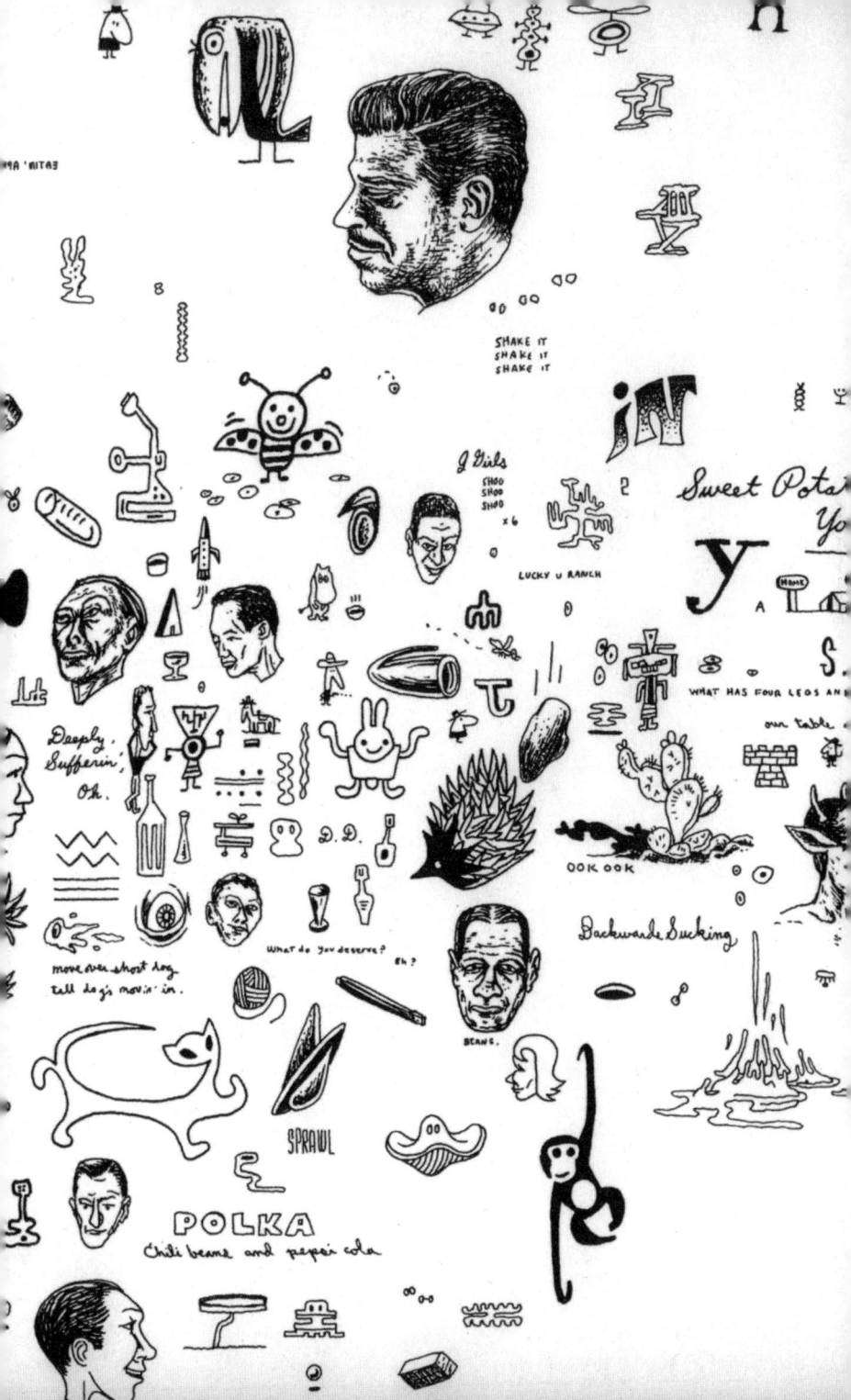

CLEAR HAT

HA

mommy

MO

L'ih

Christian Schumann 221

Lisa Oppenheim

Panorama, New York

New York, New York. Bootblack, Fourteenth Street and Eighth Avenue.

New York, New York. Statue of Garibaldi on
Washington Square.

New York, New York. Surrealistic window display, Bergdorf-Goodman.

Lisa Oppenheim 229

New York, New York. Ice skating in Rockefeller Center.

Lohren Green

from *Poetical Dictionary*

bulwark- forge into awkwardness: "bul′wärk" *n.* [from Middle Dutch *bolwerc*
or Middle Low German *bolwerk, bolle,* plank or tree trunk + *werc* or *werk,* work]

1. a broad bulky inert Being in the way, of earth or Something left stubborn,
   short to Let defenders' fire, a hard to Clamber tree-trunks and Lop stack uneven or
Lengthwise tangled in Unwieldy, splintered hatch, a mute Side, frustrating attack, Refractory.

A. a dogged Mole, sea-Wall, or breakwater, a set obstinate Crouch in currents' Weights
   and rigid, where too heavy water Breaks Itself on the Bluntness of Heaped stones.

2. *Wg.* often as Praise, a thing or person obtuse to deflection, a one Plainly staunch.

3. a planked above ship's deck Jut stiff against Sea pounding splash and wind.

contraption- through plates push a piston: "kən trap'shən" *n.*
[derivation various, thought to be a welding of contrivance with invention]
1. a half-thought scheme latched
  onto a mechanism that runs wobbling along its articulations to compile
in the course of being pushed by an often
Disproportionately Momentous
   Drive of Insistent-Centripetal Torque
     That Is Mightily Unhurled
    From The Motive Center With
      All The Myopia Of Force
and transitioned along a smoothly compliant stretch of belt
  then sped round a pulley crown
 that perforates regularly a cycling length of
stoccattically distributed cylindrical grips
    to return about a big turning
  indenturated sprocket
  that describes from the oblong back rollings and forth
of its hinged, horizontal tangent arm the
  wiper-sweep, handle-motion of a
    punctual ratchet
troping   turn by   turn by turn a
stripped bolt head
         Loading
       Loading   in the compress-spring
an eccentrically accumulative contraction
to attain a simple, finally defined, end.

that- flick an accurate tongue dart: "that" *pron., adj., adv.,* and *conj.*
[Old English *þæt*, neutral demonstrative pronoun and definite article]
I. a most general specification *that*
 is always particular,
as in *pron.*—
A. *that,*
I mean that right back there, B. *that* which is not this, or C.
that further away in space or time which
is not this,
or D. a subject or object *that,*
serves in a relative clause, or E. the convergence
of rapier and wit, take *that*-
or *adj.*— F. *that* which was already mentioned, or *adv.*—
G. to *that* certain extent it qualifies adjectives of magnitude
or *conj.*— H. *that* it might
introduce a subordinate clause as the subject
is possible or, I. tone-judging *that* person or thing
or *n.*— J. the *that* of thatness, being the particular way *that* is
necessarily what *that* is, or *pron.*— K. the attuned re-marking
yes, (all) *that.*

zoological garden- send quick strands
into twisting forms "zoological",
and set in bowls of earth "garden":
"zō′əloj′i kəl gär′d²n" *n.*
[*zôion*, animal + *logìa*, from *lógos*, logic, idea, or word,
+ garden, akin to Old High German *gart*, enclosure]
I. a reason wends its way among all
the clustered chatter and lurches, the
feedings, births, and yawns
of long-evolved beings like
monkeys maniac in their own intelligence that
fly and swing above color-coded maps
with their names in New Latin
while the hippo barely moves, its girth
sunk in a muddy likeness
of native habitat, its gray sense organs
budded open above water, culling
indifferently the daily signs of kids and
macaws and a miniature train that runs
past busy pink flamingoes who crane and
peer and preen atop stick legs,
their constant minor flaps
nervously touching a live scatterplot
of activity into their forgetful puddle,
and so from the rain forest to
the river otters to the aviary
the display of life is designed to let be
life's variable rhythms, each
animal being one other living figure
within the constraints of a composition
where everything is somehow human,
and yet everything, in its way, itself.

Our ancestors' archive rooms were filled with books, records and written accounts of their achievements while in office. Outside the houses and round the door-lintels were other likenesses of those remarkable men. Spoils taken from the enemy were fastened to their doors and not even a subsequent purchaser of the house was allowed to take these down. Consequently, as they changed owners, houses celebrated an ongoing triumph.

—Pliny the Elder, *Natural History*

John Morris

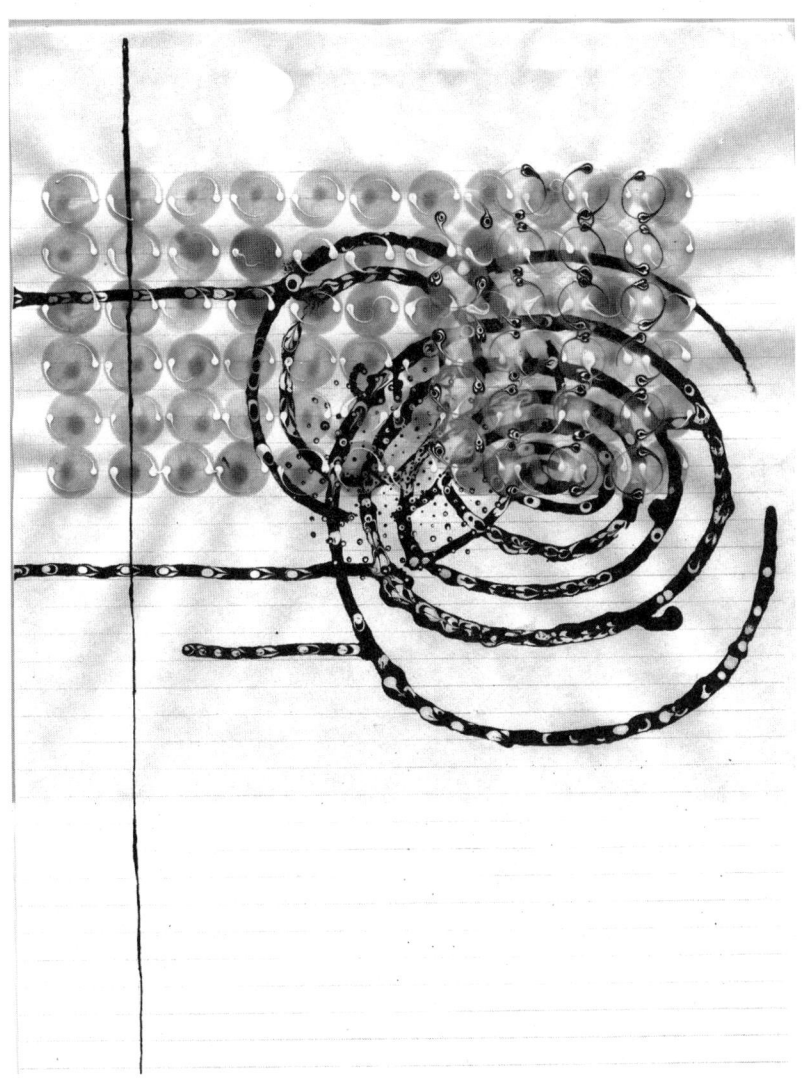

John Morris 241

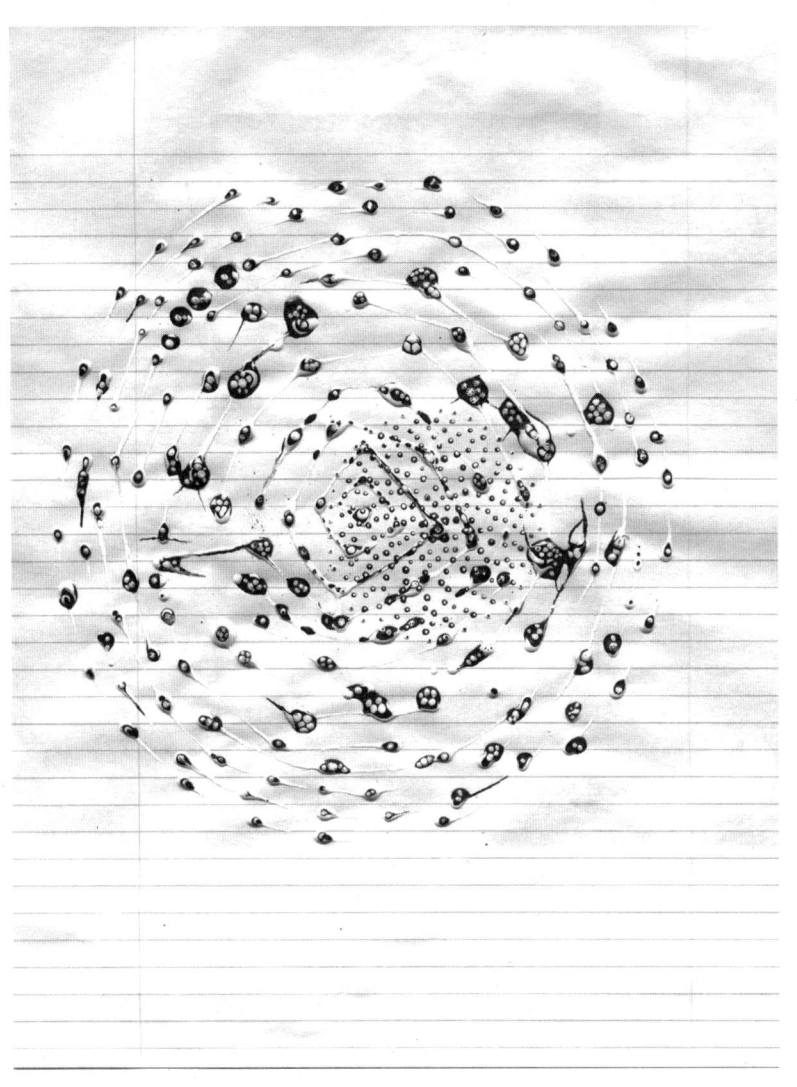

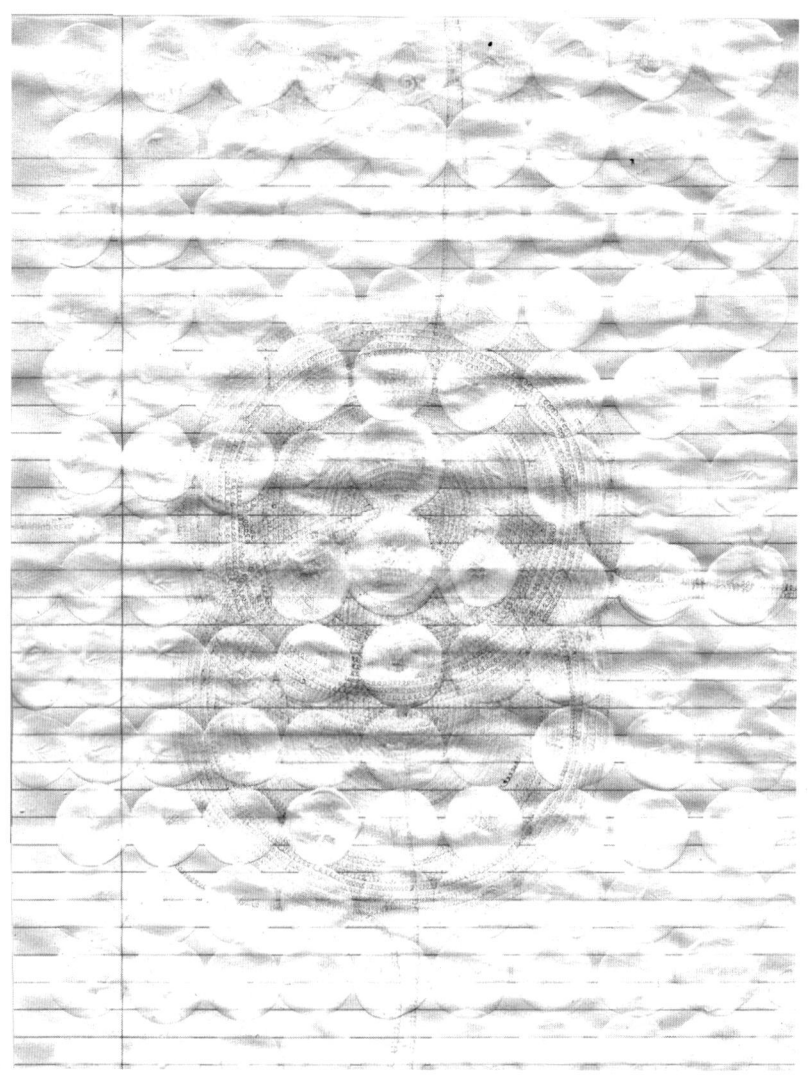

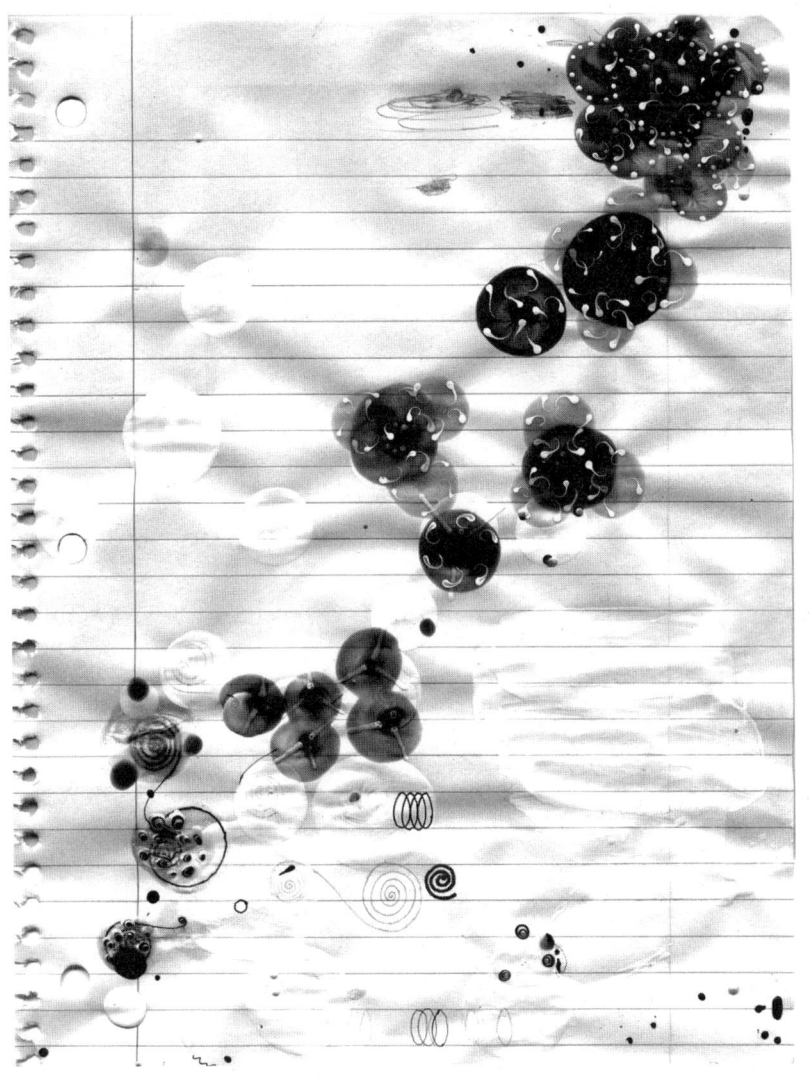

John Morris 245

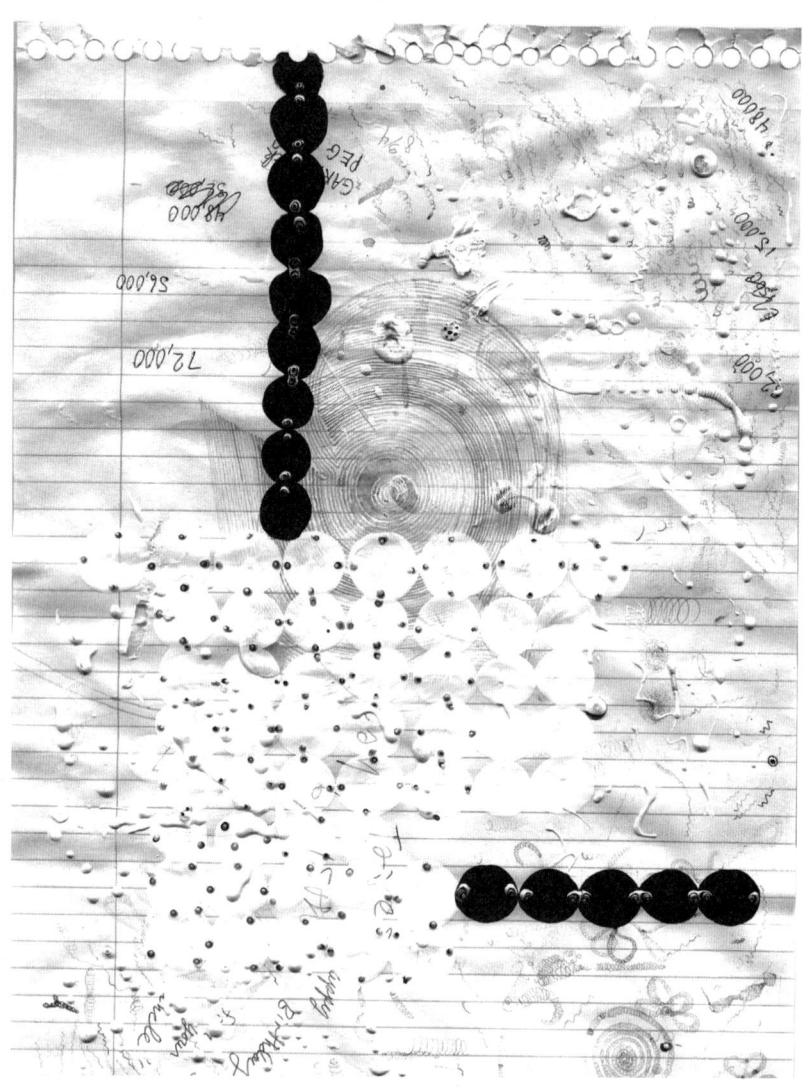

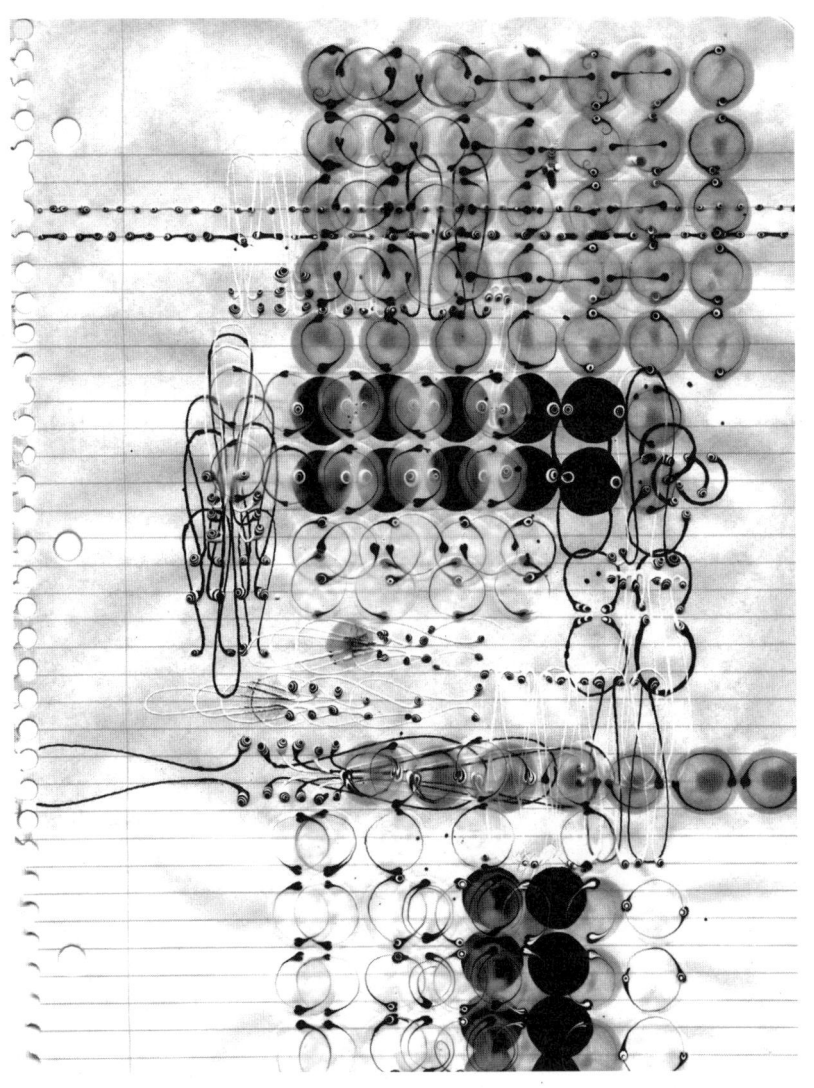

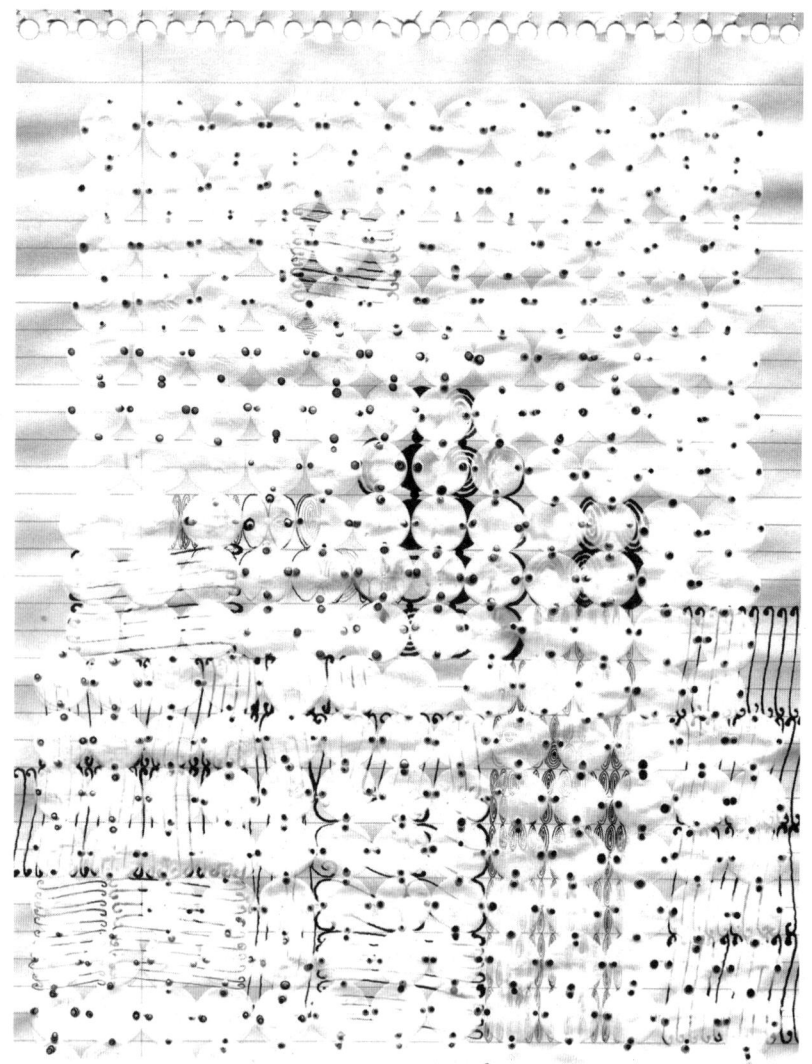

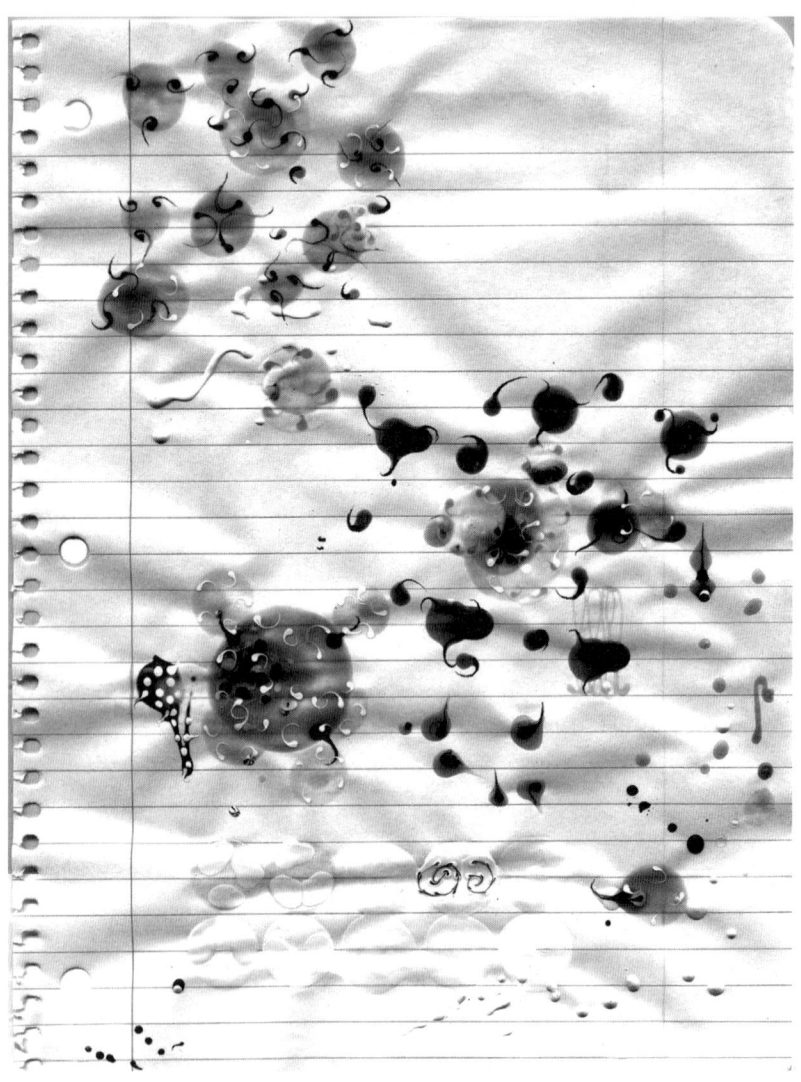

250 SHARK

Michael Scharf

Our Recognizance

Dan Farrell, Last Instance, Krupskaya, 1999
Dan Farrell, The Inkblot Record, Coach House, 2000

Dan Farrell's *ape,* written in 1988, is a serial work consisting of 20 pages divided, in Jack Spicer's *Heads of the Town* fashion, into blocks of verse-like near-lists on the uppers, and commentary-like prose on the lowers, bookended by two straight, single-line lists. Reassessing *ape* in the "Disgust and Overdetermination" issue of *Open Letter*, Farrell noted that the list items are "…words used to help people describe the pain they were in to their physicians. Annoying to Wretched, not in that order. The original diagnostic tool presented lists of words in twenty categories, divorced from any definition or context except type and intensity." The appearance of *The Inkblot Record* more than 10 years later reveals a striking consistency in Farrell's choice of source text, as *Inkblot* draws on the results of a related instrument, the Rorschach test. The book consists entirely of actual (altered?), alphabetized, unlineated, unparagraphed responses to (absent) images, separated by full stops: "Birds. Birds perched, wings out, up. Black hair strewn all over the barber's floor. Black Sabbath." While Farrell now finds that *ape* reflects a "mistake" about language's function (more on this in a minute) there is obviously an important continuity between these earlier and later books, indirectly yet also discernibly reflected in the varyingly constructed *Last Instance*.

As Farrell further remarked on *ape*'s lists: "I was struck (struck? sharply? hardly?) at the time by how these words are, were, also often used in situations other than reporting pain, that is, everywhere else. Their 'private' use, pointing to something had only by one, paralleled a 'public' one." The private language argument terminology here is wholly intentional. (Farrell's epigraph in the *Open Letter* piece—*"This? What!?"*—is attributed to Wittgenstein.) The argument against a private language, famously, hinges on reports of pain. The *Investigations* interrogate the idea of pain as a paradigmatically 'inner' phenomenon, one given its meaning by one's own experience of it, which then (as this straw line of thinking goes) directly results in the pieces of language of the kind Farrell realigns. But that inner process by which one's own experience "gives meaning" to an expression is found by Wittgenstein to be finally illusory. Expressions of pain, far from being paradigmatically private, turn out to work the same way as the rest of language: through desire to communicate, prior acquisition of shared code, reasonable guess as to intention. Meaning must always be, then, in a sense, relational: constituted in real time and dependent on social eros. It's here that Farrell locates *ape*'s error:

> my mistake then was to believe that this vocabulary was referring to experiential objects, pointing to the pain; and could also be used to describe events in the shared world. Today I am more inclined to consider language, even in its descriptive use, to be an activity uncompelled by any object. So that when I tell you my finger is Numb, or September is Flickering, it is less a description of reality than an appeal for a particular kind of attention or response. When I say appeal I also mean demand, wish, plea; clearing my throat etc.

While the private language argument had been appropriated to death even by 1988, Farrell's is a wry, knowing re-appropriation, not least because he codes the text with asides about the poetic struggle for self-definition in the face of older definers whose hands, especially then, may have seemed to be stretching directly toward one's linguistic pockets: "Lost generation, wax figure, going about it the wrong way. Not what's in my wallet." Mistake or not, his reappropriation does a lot of work when set loose on systematized therapeutic language, producing, for instance, the demand for a response, for action. And I think *The Inkblot Record* and *Last Instance* are two brilliant and very different attempts at reexamining and redressing *ape*'s demands. Farrell's refusal to accept the circumscription of psychological discourses claims a form of agency. His tactic is to set up structures for displaying sentences, and tracking their fall in and out of discourse—either as found (*Inkblot*) or (mostly) composed (*Last Instance*)—so that their affiliations with power, if not dissolve, become nuanced, i.e. regain their overdetermination. In *Inkblot*, the direct target is the huge, coercive, outdated, state-appropriated, version of psychiatry at which *ape* hints. In *Last Instance*, it is twelve diffuse, stultifying forms taken by conventionally mediated relationships—familial, significant-otherial, occupational, temporal (see the calendric send-up "366, 1996" which ends the book), etc.—picked-off one at a time at appropriately honed scales. Both books find people behind the lexical curtains.

In *Inkblot*, the aesthetic value of the Rorschach responses easily outruns whatever normative diagnostic component might remain in them.[1] In 1921, Swiss psychiatrist Hermann Rorschach published *Psychodiagnostik*, a monograph containing the blots still in use, which he developed to streamline the process of assessing personality traits and diagnosing psychosis. Having a standard set of cards, questions, and evaluative parameters, it was thought, allowed psychiatrists to learn a single process and simply repeat it, rather than relying on impressionistic and unreplicable methods, and to develop a control group of normal responses. Rorschach died the next year, but the method was eventually adapted by state and corporate entities for weeding out the bearers of potential problem personality traits, or for confirming suspicions of the possession of such traits.

The major impression left by *Inkblot* is that, lacking an aesthetics, the test, like the larger psychiatric apparatus of which it is a part, could not account for the incredible charge, diversity and beauty (even in their stretches of banality) of the responses, and thus for the people that made them. The responses are great ("If the robe was closed, you couldn't see the people or feet because the robe goes to the floor") and the abecedarian approach perfectly highlights the scale: there are thousands of subjects out there who produced these texts; a more democratic art cannot be imagined. Yet each of these responses was produced within highly alienating circumstances, which we also experience in reading them one after another, if with an almost quaint-feeling historical displacement. (Indeed, the Rorschach's low-tech lack of invasiveness in the era of Prozac, Paxil, and Nardil reads like relative humanity.) Though no longer "state of the art," the structure and content of these responses, like cries of pain (which in many ways they are), continue to claim agency *in spite of* the structure that "produced" or elicited them. There is no private language; the point is to respond, to reconfigure relations. The (I think new) form of historical consciousness the book brings forth, pitching back and forth between multi- and univocality, is a step in one possible process.

So while there is surely a critique of the Rorschach process's dehumanizing aspects implicit in Farrell's stacking up of its detritus, there are also intimations of interlocutor, of the possibilities of real exchange. The back cover of the book lists "trigger" questions, some simple ("What do you see?"), some follow-up ("What in the card gave you the impression of mice?") and others more complex, and allowing, in classic style, for a direct importation of (and, goal-wise, reconnection with) personal history ("What might that reflect in your life?"). There are also extra-linguistic responses that are included in the body of the text as a kind of wry duty, parenthetical scene-setters attributable to the original data collector: "(inverts card)," "(imitates sitting)," "(traces outline with finger)," "(demonstrates the sun's rays using her body)," "(pantomimes stomping)," "(laughs)," "(runs finger around blot)." Phatic openings play a huge organizing role here—many, many responses begin with "Well..." "Oh..." "Um...."—which is certainly comic, but also points to real people involved in real speech events, talking to others. All of this is data which, by the 50s and 60s, actually would've been taken into account by someone administering the test, as evaluation moved away from just the contentful construal of imagery. By silently moving these aspects of the data to the fore, as the result of a further (abecedarian) "procedure," Farrell simultaneously makes us register the protocols of a system that attempted to record and evaluate even the smallest indirect response—a system of coercion and control—as well as the possibilities of a modest recuperation, of a provisional re-interpretation.

A fantasy-driven aside along those lines about source text-driven work: Beyond the classificatory grid that presupposed a set of typologies for quick mid-century "scoring" by prospective employers or wardens, one can imagine a therapeutic component to the test. Rorschach began his career as an advocate for the then fledgling psychoanalysis. Though I haven't read *Psychodiagnostik*, it's possible to believe that some of the therapeutic aspects of dream interpretation or word association made their way into its design, at least initially—that further exchanges that were supposed to "happen" during Rorschach tests along the lines of Freudian psychiatry, analogous to changes in the self that were (are) supposed to occur when conceptual connections, were spontaneously generated during an analytic session. A similar thing supposedly happens, one line of thought goes, in the composition of poetry: the composed text is the record of the poet's own achievement of spontaneous connection, which simultaneously produces and is produced by platonic ecstasy, which readers follow and try to reproduce via the poet's traces. Many readers, perhaps unconsciously, still put source-based work in a different, less ecstatic category. PET scans might reveal that there are no differences in the brains of those re-arranging pre-existing orthographic symbols vs. those generating them *ex nihilo*, so we'll still have to wait for the data on whether the *bricoleur* and text collector are in fact "real artists" by that definition. But *Inkblot* allows the thought that the generation of the Rorschach responses might have changed the subjects' psyches directly—that actual therapeutic "work" got done, even if it wasn't usually considered the main point of the test. *Inkblot*, divorcing the material of the test from its supporting superstructure, shows that its own banal organization scheme clearly does as much psychic work for us, if not more, than the responses-as-text. All traces can be equal. Any found work makes a similar statement, but *Inkblot's* material, read as therapeutic, heightens the contradictions definitively.

If *The Inkblot Record* records through a kind of one-way mirror, then *Last Instance* tracks the meditative streams that lie behind the design and construction of the glass, and the modes and means of contact through it: "For years call to pretend calling, each other a stake in spontaneity. And what of it? How can I be but thankful for the boring contact. It would only be unhelpful to remind them this is actually a late returned call, not of their own accord. I ignore the slight. It would be selfish of me not to pretend." These lines are from (the by now just-ly well-known) "K," which fugues (I think another reviewer has said this) around the possibility of contact with the eponymous character, and throws off a profusion of nuanced local reflections about the nature of supposed intimacy. In so doing it illustrates how we, "callers" and "callees" all ("I am the caller I have always been"), are forced to project ourselves onto nothing, rats who press the buttons when the empty light of desire goes on. Talking to a machine (and later imag-ining how K's childhood produced the class-based eschewal of man-ners that may really be the key to happiness) or actually getting the roommate are the same thing. The protocols of contact replace the contact itself.

Similarly, "Avail" takes, permutes and distorts seeming multi-ple choice answers to a sort of health questionnaire, revealing the sutures where corporate State enters subject as host. Staged from deep within the logic of the medical-industrial complex—that confluence of your job, your insurance, your health care provider, your self-image, your home environment, and advertising—the piece performs a bril-liant reduction of the system's terms to "anger" and "physical health," whereby the seething indignation that lack of self-determination pro-duces is re-assigned back to the subject in an infinite recursive loop: "I feel like my physical health is something that I myself am in charge of. I feel angry about myself a good deal of the time. Things are not more irritating to me now than usual...I'm very direct with people when it comes to my own physical health needs. When I become sick or ill, I am the person to blame. Sometimes I am so angry that I feel like hurting others, but I would not really do it." The "I" here is sim-ply a pivot point for contradiction. The banalization of health, anger and violence, emptied of anything other than their significance as claims on resources and their inability to be reconciled, is an integral part of the system's self-representation, as if a built-in resistance to articulation, let alone resolution, were a normal part of its proper functioning. Seem familiar? It's a poem that should be printed on the back of every Oxford Freedom plan.

The book's five shorter pieces ("Stillstand"; "Fuming"; "Invigorator"; "Geal"; "Jumb") track "particle[s] of a universal waning of a class" with "no redreaming qualities," and the algorithms that hold it in place: "Let's say: 'X.' *Now that is going too far.*" Yet there are movements toward escape. Following the verse alterations of 1994's *Thimking of You*, which quietly unseated the word ("F/ ear is my master"), *Last Instance*'s prose sentences hold the axis of combination steady while radically compounding that of selection, and thus demonstrating the possibility of movement even within normative convention. Its quick parodies of cliché—"Delease me" "Stop melee I'm robot to burst" or "Out of the radish into the pan" (which has neither frying nor fire, yet both are there)—are followed up by straightforward descriptions, which direct our heightened attention to finely resolved grammar-pictures that can then seem to point beyond their ostensive objects: "Pig iron on the anvil cooled misshapen, steaming coffee on the burner burned." Or "The cat communicates by doing nothing. The feathers are warmest. Even the climate is huffy. Let's dip in to light the candles. The coin is in the bucket and the earnings are battered ready to drop." Every expression of stasis is also a wry demand for attention to the possibilities, however compromised, that remain.

That aspect of the book comes through most clearly in "My Recognizance," the longest poem in the book. "Recognizance" seems like a Farrellian neologism, but my dictionary lists the following definitions:

> **re cog'ni zance**, n. [OF. reconoissance, later recognoissance, deriv. of L. recognoscere, fr. re- + cognoscere to know.] 1. Law. An obligation of record entered into before some court or magistrate, making the performance of some act the condition of nonforfeiture. Also, the sum liable to forfeiture upon such an obligation. 2. Archaic. A token; symbol; pledge; badge.

The poem is thus a reckoning and recording, in which the poet both recognizes and reconnoiters his childhood, adolescence and early adulthood with acceptance but without intending to linger nostalgically, "Away, a moment."

While outside in crowded cards of skilled hockey players I

saw my own reeling life clasped and slipped to clipping spokes. Sinking under shuttered glands, chafing beneath my dolled tonsils, with no effort I wafted upon my mother's brownie effs. I cannot see the jaguar for its spots, tottering there above the fizzling bones, the tiny gristle, the flickering marrow. Pups of twisted balloons yipped by. Was lemon, salmon, or yellow, membrane fixed, as a bicyclist to her pedals, to the corollal lambskin skin. Swaying, seen in the breeze of winterly floss a quantum of batting was puffed away. A melt the hot toy plastic descended, in droplet to me. I could not force these primary emotions, like a lunch, sandwiched between definition and extension, on me. In decrements imconpetent, nodding off during inattention spells, disguising the shallow branch tracks with slight foot-marks. What is a modificationist today may be a hoofer tomorrow.

There's a pun on spokes in the first sentence, where biking and speaking, both become, sort of, jokingly, castrating ("clipping")— a kind of capture, "shuttered," as one is by maternal photography ("my mother's brownie effs"), which unfailingly affixes an image of the child's endearing awkward stage, glandularly produced. The toy images of living things we interact with in youth are not analogues— they *are* life, those pups really yip. The "lambskin," is a jacket—or a condom, a false flowery crown ("corollal") that gets "fixed" to us, sex, since we're not fixed. Intellectualizations ("definition") and things themselves (philosophical "extension") often add up to the same distasteful coercions, some of which we force on ourselves. Distinguishing these things is disgusting, and like death (there is "excrement" and "cremate" in "decrement," an obvious "con" in "imconpetent," and a retch in "force"), either at the time or in remembering, and we're not always up to it, and we cover it up, "disguising" the tracks with "slight foot-marks" (I'll say it: writing). What we might try to deal with at one moment we jauntily lie about or run away from the next, but the attempt, however tongue-in-cheek, to construct a usable past is inevitable. We are obliged to record the past, or its specifics simply recede into the general hot plastic—i.e. are liable to forfeiture. And we may even derive some pleasure from it.

"My Recognizance" doesn't end there, and the above passage for it is somewhat unfair to isolate and try to close read or force allusions on, since it uses the most transparently autobiographical language in the book. But Farrell is able to construct such passages so that it seems unnecessary to try to attach them back to a "speaker" of the poem, or to the author function—rather we are able to examine such specifics of a life as if laid out on an archaeologist's sorting table; they are the "fizzling bones, the tiny gristle, the flickering marrow" from which we are to project the general species.

So if this is a Beckett-like move, as has been suggested, it's the Beckett where characters manage small pockets of self-determination, like the rigorous organization of one's oral fixation, or the endless recontextualization of a lexeme, like "on" in *Ill Seen Ill Said*. The kind of intentional lightness that pervades is anathema to ritual suckings and pocketings (though it would not reject them), and the non-neurotic attention that results is a model. It's free and it's available, and it has other precedential takers-up: the opening of *Last Instance*'s first piece, the wryly pistolized "No Future," riffs simultaneously on late Beckett's relentless self-reflexivity, Silliman's adaptation of it in his opening tjant, and Wittgenstein's rejection of self-ostentive privacy:

> The opposite up, the opposite early, fast put back, kaput. Had gone to the said beginning, again had gone. Now what. Put back and begun. Head. Kaput. Present perfect. But had begun. Again gone again more. Then lest then again. The snow mountained up, stopped, had it finished. As in, up, now what. Kaput. Now what.

Wittgenstein: "This?—what?!" Beckett: "It stands. What?" Silliman: "Not this. What then." Farrell: "Now what." Both of Farrell's recent books evince care at every decision point, and a drive toward negotiable social meaning. In that they work to reorient us within their moment, whether just ended, or still evolving. The idea is recognizance: to reconnoiter, recognize, reckon, rethink, record and respond.

Note

1. And remain it does: check out www.deltabravo.net/custody/rorschach.htm for tips on beating the standard 10-card test, often administered as part of custody battles.

Notes on Contributors

Jimbo Blachly is an artist living and working in NYC. He will be having a solo exhibition of his work at the Sculpture Center in New York in December 2002. He is also working on *Specific*, a collaboration with Lytle Shaw that was exhibited as a work in progress at ACE Gallery in May 2002. **Olivier Brossard** is currently writing a Ph.D. dissertation on Frank O'Hara (University of Paris 7). He is co-editing an anthology of poetry in French with Marcella Durand and Kristin Prevallet for Talisman and is one of the editors of the French and American website www.doublechange.com. **Matthew Buckingham** is an artist living in New York. His recent book *Subcutaneous* was co-published by Shark Books and Murray Guy Gallery. **Paul Chan** mwgww.nationalphilistine.comansgh j2hsk. **Andrew Clark** will begin an assistant professorship in 18th-century literature at Fordham University this fall. He works on aesthetics, physiology, and music in the 18th century and is currently finishing his dissertation: *Diderot: The Poetics of Physiology*. **Brandon Downing**'s volumes of verse include *Lazio* and *Dog and Horsey Pictures* (both Blue Books, 2000) as well as *The Shirt Weapon* (Germ, 2002). Downing lives in New York City. **Rob Fitterman**'s poetry books include *among the cynics* (Singing Horse, 1991), *Metropolis 1-15* (Sun and Moon, 2000) and *Metropolis 16-29* (Coach House, 2002). Fitterman currently lives in Florence. **Lohren Green** is a writer living in San Francisco. He's currently finishing a book-length manuscript of poetry, prose, tables and charts entitled *Poetical Dictionary*. He has a Ph.D. in rhetoric from U.C. Berkeley. **Yun-Fei Ji's** work was recently featured in the Whitney Biannual 2002. He shows his work with Pierogi, Brooklyn, where he had a solo show in 2001. He will be having a solo exhibition at the St. Louis Museum of Contemporary Art in the fall of 2003. **Nina Katchadourian** exhibits with Debs & Co. gallery in New York and Catharine Clark gallery in San Francisco. She currently lives in Brooklyn and teaches at Brown University. **Matt King** is a sculptor who's work has shown in New York, Texas and Thailand. In 1999 he had a solo exhibition titled Mergers at Audeillo Fine Art in New York City where he lives and works. **Tan Lin**'s poetry books include *Lotion Bullwhip Giraffe* (Sun and Moon, 1996). Two books are forthcoming: *Kruder & Dorfmeister*, a novel in photographs (Faux Press) and *IDM* (Atelos). Lin lives in New York City. **Carl Linnaeus**'s scientific works include *Musa Cliffortiana* [Clifford's Banana] (Leyden, 1736), and *Viridiarum Cliffortianum* [Clifford's Pleasure Garden] (Amsterdam, 1737), as well as several travel accounts commissioned by the Swedish Crown. He was professor of science in Uppsala from 1741 until his death in 1778. **Pamela Lu** is author of *Pamela: A Novel. A Classification of the Minor Transformations: Category Am* is forthcoming from Shark. Lu lives in Los Gatos, California. **Bernadette Mayer**'s numerous books of poetry include *The Desires of Mothers to Please Others in Letters* (Hard Press, 1994) and *Another Smashed Pinecone* (Granary, 1998). She lives in East Nassau, New York. **Scott McCarney** is an artist and designer based in Rochester, New York, whose art has found its home in the book form for over twenty years. Many of his one-of-a-kind books, offset and small editions, and sculptural installations make use of discarded encyclopedias rendered obsolete by the ravages of time. **Ange Mlinko** is author of *Matinées* (Zoland, 1999) and was recently editor of *Poetry Project Newsletter*. She lives in New York. **John Morris** is an artist living and working in Queens. His drawings were recently included in the Museum of Modern Art's exhibition *New to the Modern: Recent Acquisitions from the Department of Drawings*. He will have a solo exhibition in September 2002 at D'Amelio Terras Gallery, New York, where he is represented. **Tue Andersen Nexø** lives in Copenhagen, Denmark, where he studies comparative literature. He has published critical essays on contemporary Danish poetry and co-edits the literary magazine *Den Blaa Port* (The Blue Gate). He co-organised *In the Making*, a Danish-North American poetry conference. **Novalis**'s poetry has appeared in *Athenaeum* 3:2 (1800). He is also author of the critical work *Pollen* (Freiberg, 1798; later re-titled *Miscellaneous Observations*). *General Draft* was written over the winter of 1798-1799. **Redell Olsen**'s recent work has appeared in *Parataxis, How (2)* and *The Paper. Her Book of the Fur* was published by rem press (Cambridge) in 2000. **Lisa Oppenheim** is an artist living and working in New York City. In addition to his 37 volume *National History* (private scribes, AD 79), **Pliny the Elder**'s works include *De iaculatione equestri*, a treatise on the use of the javelin as a cavalry weapon (private scribes, AD 55?). Born in Novum Comum in 23, he was overcome by sulphurous fumes while observing the eruption of Vesuvius in 79. **Michael Scharf** is the author of *Telemachiad*, and editor of Harry Tankoos Books. He lives in New York City. **Jovi Schnell** is an artist living and working in Brooklyn. She exhibits with Derek Eller in New York City. **Susan Schultz** teaches English at the University of Hawai`i. She is author of *Memory Cards & Adoption Papers* (Poets & Poets, 2002) and *Aleatory Allegories* (Salt, 2000), and edited *The Tribe of John: Ashbery and Contemporary Poetry* (Alabama, 1995). She edits *Tinfish*, a journal of experimental poetry from the Pacific. **Christian Schumann** is an artist currently living in

Brooklyn. An exhibit of his work is scheduled for the fall of 2002 at Postmasters gallery in New York City. **Brian Kim Stefans** is the author of three books of poems, the most recent of which is *Angry Penguins*. He is editor of the website Arras: new media poetry and poetics. *Fashionable Noise: On Digital Poetics* (Atelos) is due to be published in fall 2002. **Juliana Spahr** is the author most recently of *Fuck You-Aloha-I Love You* (Wesleyan, 2001) and *Everybody's Autonomy* (Alabama, 2001). **Heriberto Yepez** (born Tijuana, Mexico) lives at the border and has published essays, short stories and poetry in Mexico, and now is writing and doing readings in English. hyepez@hotmail.com